建筑师标准信函格式（原著第三版）

［英］大卫·查贝尔 著
卢昀伟 王贤芬
张男 李东 刘军 译

中国建筑工业出版社

著作权合同登记图字：01-2005-5998 号

图书在版编目（CIP）数据

建筑师标准信函格式（原著第三版）/（英）查贝尔著；卢昫伟等译. —北京：中国建筑工业出版社，2006
ISBN 7-112-08724-4

Ⅰ.建... Ⅱ.①查...②卢... Ⅲ.建筑工程—信函—范文 Ⅳ.TU723.1

中国版本图书馆 CIP 数据核字（2006）第 121513 号

译自“David Chappell：Standard Letters in Architectural Practice，3rd edition”

This edition is published by arrangement with Blackwell Publishing Ltd，Oxford.
Translated by China Architecture & Building Press from the original English language version. Responsibility of the accuracy of the translation rests solely with the China Architecture & Building Press and is not the responsibility of Blackwell Publishing Ltd.
本书的翻译出版得到英国 Blackwell Publishing Ltd，Oxford 的大力支持。

责任编辑：率 琦 丁洪良
责任设计：郑秋菊
责任校对：张树梅 王雪竹

建筑师标准信函格式（原著第三版）
[英] 大卫·查贝尔 著
卢昫伟 王贤芬 张男 李东 刘军 译

*

中国建筑工业出版社出版、发行（北京西郊百万庄）
新 华 书 店 经 销
北京嘉泰利德公司制版
世界知识印刷厂印刷

*

开本：787×1092 毫米 1/16 印张：27 字数：657 千字
2006 年 11 月第一版 2006 年 11 月第一次印刷
定价：**59.00** 元
ISBN 7-112-08724-4
(15388)

本社网址：http://www.cabp.com.cn
网上书店：http://www.china-building.com.cn

译者序

中国自从2001年加入WTO，至今已届五年之约。我国入世时关于建筑领域勘察设计行业的承诺是：中国加入WTO五年后，允许成立外商独资设计企业。因此，可以以2006年和2007年为一分水岭，以后将会有更多的外籍建筑设计机构以独立身份进入我国建筑设计市场，随之大批的外籍建筑师也将与我国相关部门及个人展开方方面面的联系。另一方面，随着我国的建筑承包领域逐渐与世界并轨并积极参与国际工程项目的建设，自然也会经常与外籍建筑师进行沟通联系。

我国的建筑领域，特别是设计行业若想与外籍建筑机构在国内的竞争中得以保全，尤其是希望能够在未来的国际舞台上一展雄风的话，就要完全了解国际通行的执业“法则”和学会执业沟通的方法，也就是说在例行公事方面，外籍建筑师要怎么做你也必须怎么做。也许未来阻碍我们跨出国门最重要的因素恰恰就是国际执业“法则”和沟通方法。在欧美，建筑师充当的角色更加丰富，就如同委托人的全权代表，需要很强的沟通能力和迅速做出反应的能力。在当今国际惯例中，建筑师在其执业过程中与业主（即委托人）、政府相关职能部门以及承包商等之间的沟通联系主要依靠频繁的公函往来，这是一项十分重要而又繁琐的信息管理工作。作为业内人士，我深深知道国内的建筑师同行往往都缺乏这方面的沟通能力与常识，至于国际上通行的建筑师日常执业公函对他们来说更无异于天方夜谭。所以，这一课一定要尽快补上。

英国建筑专家大卫·查贝尔先生编著的这本《建筑师标准信函格式》，包罗了建筑师在其执业全过程中可能出现的常见情况下应撰写的各类信函，共计285封。这些信函格式正确、语言规范、用词妥切，现以中英文对照的形式出版，它是我国建筑师及建筑行业其他人员难得的一本好书和实用手册。在中国建筑工业出版社的大力支持下，本书的出版必将对我国建筑设计机构，参与国际工程项目的专业机构和从业人士带来极大的帮助，同时它也是我国高等院校建筑专业教学的一本很好的辅导读物。

本书在引进、翻译、编辑和出版的过程中，得到了有关方面和人士的大力协助与支持，本中心特此致以衷心的感谢。这里尤其要感谢中国建筑设计研究院张男建筑师、天津大学李东博士和北京建筑设计研究院刘军建筑师的鼎力加盟。还要感谢中国建筑工业出版社有关编辑的辛勤劳动。有了他们的共同努力，此书才得以早日面世。

最后希望各界朋友能够对本书提出宝贵的意见与建议。

卢昫伟
中外建科文化发展中心
2006年7月13日

原著第三版序

本书的上一版本获得了很大的成功。有几次我在别人请我审校的文稿中看到了一些信件，居然发现其中有按照我编写的这本《建筑师标准信函格式》所写的信函，心中感到十分欣慰。这些信手拈来的证据佐证了本书的实用性，也证明了我辛苦的收编工作是非常值得的。

我对上一版本中的所有信函进行了审校，并根据国家的判例法与成文法增删了一些信函。作为英国皇家建筑师协会（RIBA）的专业顾问，我解答过数以千计的问题，因此根据自己积累的经验对本书作出了一些改动。本书目前收录了285封信函。除了2003年3月31日实行的标准施工合同格式（JCT98）修正案外，本书还配套使用了标准施工合同格式（JCT98）、承包商负责设计的标准施工合同格式（WCD98）、中等规模标准施工合同格式（IFC98）、小型建筑工程协议（MW98）以及房屋建筑与土木工程标准施工合同格式的一般条件〔GC/Works/1（1998）〕等的最新版本。本版书参考了新的1999年版和2000年版的英国皇家建筑师协会建筑师委托协议标准格式（SFA/99），从而替代了1992年的旧版本。此外，本书还引入了建筑师委任协议（CE/99）和小型建筑建筑师委任协议（SW/99）作为参考。

我对于条款参考部分进行了必要的修改。每章前面解释说明的文字仍保持最少篇幅，希望信函自身及其标题和注释能够不释自明。经常有人对使用标准信函大加批评，认为这样会使人懒于思考并导致错误频生。但是我却认为使用标准信函完全必要，因为这样能减少工作的乏味性，使忙碌的从业者们能有更多的时间去完成更重要、更难写的信函。

最后感谢我的夫人玛格丽特一如既往的支持。

大卫·查贝尔 David Chappell
于英国塔德卡斯特 Tadcaster

目 录

绪论

绪论

本书是为建筑师所写的，但建筑行业的其他人员亦可从中获益。本书由一系列标准信函组成，这些信函往往需要反复撰写。本书试图包罗建筑师在项目处理中遇到的所有常见的情况。当然，要做到这一点难度很大，有些情况难免会挂一漏万。作者非常欢迎读者通过出版商就标准信函提出宝贵的建议，以便今后再版时予以补充。

有了这些信函，并不意味着读者就不需要了解合同了，它们只能免除在非常规范的情况下撰写信函的乏味工作。为了把大量的信函缩减到可以分类管理的篇幅，我把所有的信函按照英国皇家建筑师协会编写的《工作计划》分成了若干部分。值得注意的是，由于英国皇家建筑师协会的《工作计划》在 1999 年进行了修改，因而本书中的分类也相应地进行了少许调整。所有函件均按照逻辑上先后的顺序进行排列。

除非每封函件另有说明，所有函件都适用于标准施工合同格式（JCT98）、承包商负责设计的标准施工合同格式（WCD98）、中等规模标准施工合同格式（IFC98）、小型建筑工程协议（MW98）和房屋建筑与土木工程标准施工合同的一般条件〔GC/Works/1（1998）〕。当不同的合同要求撰写不同的信函时，均有注释加以说明。为了简洁明了，每封函件一般只涉及一个单独的主题，但在实际应用中，却常常倾向于在一封信函中涵盖多个主题。从其他途径能够得到的标准文件，诸如投标函标准格式和各种证书格式均未收入本书。

在使用本书时，应记住以下几点：

• 每封信函都应该有一个标题，标明项目的名称。但是，为简明扼要起见，书中的标题一概省略。

• 为方便阅读，书中一律使用男性的“他”。但“他”有时也可以用来表示“她”；同样地，“他的”也可用来表示“她的”。

• 根据不同的合同类型及其他目的，通常会在同一封信函中标出内容上的变化。但为了更方便、减少混淆或者当涉及到重要的信函时，其他的函件格式都会单独另立，并在函件编号上增加 a、b、c 等符号以示区别。

• 为保持一致性，书中通篇采用“建筑师”、“委托人”或“雇主”等称呼语，但应该指出，当采用 GC/Works/1（1998）合同时，建筑师被称作“项目经理”（PM）。采用其他合同格式时通常由建筑师所写的信函，在采用 GC/Works/1（1998）合同格式时，有时则是由雇主所写。

• 当采用 WCD98 合同格式时，“雇主代表”取代了“建筑师”这一称呼，因为建筑师通常被认为充当了“雇主代表”这一角色。在本书中如果出现太多诸如“雇主的顾问名称来回转换”这样复杂的问题，就会让人读起来难以捉摸，在任何情况下都不应该鼓励此类事情的发生。但这并不意味着部分信函在这些情况下不能适用，而是在使用标准信函时需要特别注意。

- 通常认为，补充条款应与 WCD98 合同配套使用。

应引起警觉的是：标准信函是十分有用的，但是，如果不加思考就照搬照抄，则是非常危险的。在使用标准信函时，一定要认真考虑该函件是否真正适合于所表达的特定情况。有疑问时，就应该寻求帮助。

第一章　项目评估

英国皇家建筑师协会最近对工程各阶段进行重新命名之后，项目评估阶段和基本设计要点阶段所涵盖的明确行为界定就有些含糊。以下的信函已经进行了重新整理，以体现修改后的项目评估阶段所包括的内容。但有一点很明确，即建筑师在执业过程中的做法是不尽相同的，而且这些不同之处在很大程度上取决于工程的规模和类别。项目评估阶段的信函主要是关于建筑师与委托人的初步关系、可行性研究以及棘手的担保书和职责协议等问题。

在这一阶段，建筑师要收集大量的信息，这从以下的信函中便可知一二。建筑师将初步接触包括当地规划部门在内的一些官方机构。

对于小型工程，建筑师不太可能写一份正式的报告，但有一点要记住：一份结构合理的报告书要比一封三四页长的信函好得多。布莱克威尔·萨恩斯编著的《建筑师及项目经理报告写作》（1996 年）一书第三版中提供了很多实用的报告格式，可以教你在执业过程中怎样写作既简练而又全面的报告。

建筑师确认与委托人之间协议的所有细节是非常重要的。因为这些事情是再平常不过的例行事宜，反而容易被忘记。尽管有些情况下需要撰写特定的信函，但标准信函和其他标准文件一样是很有用的，比如 1999 年的 RIBA 建筑师委任协议标准格式（2000 年版）（SFA/99）、对中型工程适用的建筑师委任协议（CE99）和对小型工程适用的小型建筑建筑师委任协议（SW/99）等。签订协议时，要认真考虑《1999 年客户合同中的不公平条款规定》对此的影响。最近颁布的判例法规表明，当委托人为独立客户，尤其当用到 RIBA 条款时，需要严格执行该规定。客户合同必须要通过各自协商，并向委托人解释清楚内容。即使是仲裁和裁决也有可能被认为是不公平的。

确保你的委托人直接雇佣所有的顾问，这是明智之举。如果顾问玩忽职守，SFA/99 中的标准条款可以保护建筑师不受追究。

没有委托人的明确授权，建筑师不可以委托他人进行工程设计。另外，建筑师要确认委托人直接与承担设计工作的分包商签订协议，以便委托人可以在分包商设计工作万一出现失误的时候得到赔偿。这些协议通常被称为“担保书”。建筑师通常需要签订一份有利于雇主或投资者的可转让的担保书。如果没有这份担保书，建筑师在设计失误的问题中实际承担的责任可能会微不足道。建筑师往往不愿意签订这种担保书，但迫于商业竞争的压力常使得他们不得不签。

函件 1

当委托人要求建筑师为设计服务报价时，致委托人

Letter 1

To client, if asked to tender on fees

尊敬的先生：

你方于［填入日期］的来函收悉，深表谢意。

得知你方拟委任我方为以上工程项目的建筑师，我方深感荣幸。RIBA 的《职业操守规范》规定了我方作为建筑师在报价之前，必须遵守其收费标准。这些是与项目的性质、规模以及所需要的服务相关的。

可否请你方与我方电话联系，以便商定合适的日期和时间来讨论此项目。随函附上一份建筑师委任协议标准格式（SFA/99）［当委托项目为中型工程时应为“建筑师委任协议（CE/99），”小型工程时应为“小型建筑建筑师委任协议（SW/99）”］作为一般的信息参考，并帮助你方明确所需要的服务。如果得到委任，我们的协议将建立在此标准格式的基础之上。

期待与你方面谈。

你忠诚的

Dear Sir

Thank you for your letter of the [*insert date*].

I am pleased to hear that you are considering my appointment as architect for the above project. The RIBA Code of Professional Conduct lays down criteria which I must satisfy before quoting a fee. These relate to the nature and scope of the project and the precise services required.

May I suggest that you telephone me to arrange a suitable date and time to discuss the project? A copy of the Standard Form of Agreement for the Appointment of an Architect (SFA/99) [*substitute 'Conditions of Engagement (CE/99)' when dealing with medium sized commissions and 'Conditions of Appointment for Small Works (SW/99)' when dealing with small commissions*] is enclosed for general information and to enable you to form an idea of the services you will require. If appointed, it will form the basis of our agreement.

I look forward to meeting you.

Yours faithfully

函件 2

致预期的委托人，提供服务

Letter 2

To prospective client，offering services

尊敬的先生：

得知你方计划进行［填入工程性质］，我方对此非常有兴趣。不知你方是否已经为此工程委任了建筑师，如果尚未委任，我方希望能担此任。如果能与你方讨论这一项目，我方将深感荣幸。

随函附上我方的插图作品宣传册，希望你方对此有兴趣。从中，你方可以看到我方对你方有意向进行的项目是非常有设计经验的。

如果你方认为面谈将对双方更加有利，请告知我方。

你忠诚的

Dear Sir

I was interested to hear that you intend to［*insert nature of development*］. I do not know whether you have already commissioned an architect for the work. If not，this letter is to let you know that I would be delighted to discuss the project with you.

A copy of my illustrated brochure is enclosed and I hope you will find it of interest. You will see that this practice is experienced in carrying out work of the kind you appear to have in mind.

If you consider that a meeting would be mutually beneficial，please let me know.

Yours faithfully

函件 3a

致委托人，说明 SFA/99 委任条款

Letter 3a

To client, setting out terms of appointment SFA/99

尊敬的先生：

谨提及［填入日期］关于委任我方为以上项目建筑师的双方谈话/你方来函［视情况取舍］。我方愿意接受这一委任，并深感荣幸。

现确认此项目将使用 RIBA 建筑师委任协议标准格式（SFA/99）中规定的条款和条件。现随函附上两份标准格式。这些条款是根据我们［填入日期］的谈话中讨论的条目完成的。如果你方能在黄色标签和铅笔十字标记所示区域内完成这份文件，并给我方寄还一份，我方将不胜感激。另外一份留于你方作为信息参考。

请核查一下所报各项服务和费用计算标准是否令你方满意。如果你方有什么疑问，请即刻与我方联系，以便给予解释。

你忠诚的

Dear Sir

I refer to our conversation/your letter [*delete as appropriate*] of the [*insert date*] regarding my appointment as architect for the above project. I am happy to accept this commission.

I confirm that the terms and conditions which will apply to the project are those set out in the RIBA Standard Form of Agreement for the Appointment of an Architect (SFA/99), two copies of which are enclosed. They are completed in accordance with the terms we discussed at our meeting on the [*insert date*]. I should be pleased if you would complete the documents in the spaces indicated by yellow stickers and pencil crosses and return one copy to me. The other copy is for your retention and information.

Please check that the services and basis of fee calculation is satisfactory. If you have any queries, please do not hesitate to ask me for clarification.

Yours faithfully

函件 3b

致委托人，说明 CE/99 委任条款

当委托人为独立客户时，此信函不适用

Letter 3b

To client, setting out terms of appointment CE/99

This letter should not be used where the client is a consumer

尊敬的先生：

谨提及［填入日期］关于委任我方为以上项目建筑师的双方谈话/你方来函［视情况删除］。我方愿意接受这一委任，并深感荣幸。

现确认此项目将使用 RIBA 建筑师委任协议（CE/99）中规定的条款和条件。现随函附上两份标准格式。这些条款是根据双方讨论的条目完成的。

请特别注意以下条款：

1. 本合同遵守［填入相关法规，如“英格兰和威尔士法律”或“北爱尔兰法律”］。
2. 在第 7.2 条中提到，本协议规定的起诉时效为［填入年数，记住非盖印合约的法定年限为 6 年，契约合同的法定年限为 12 年］。
3. 在第 7.3 条中提到，我方的责任限额为［填入你方责任限额，把你方的专业责任保险（PI）及可能的建筑成本都计算在内］。
4. 我方对单一事件或单一事件引起的连续事件投保专业责任保险，投保金额为［填入金额］。
5. 第 9.5 条规定了仲裁为解决争端的基本方法，这一规定不影响裁决权的使用。如果双方不能就仲裁人达成一致，任何一方都可以向英国皇家建筑师协会申请指定仲裁人。［如果更希望通过法律程序来解决争端，则替换为：］双方同意关于仲裁的第 9.5 条不适用，如果发生争端，任何一方都可以诉诸法律程序。
6. 本协议生效日期，即我方开始服务的日期，将为/［视情况取舍］［填入日期］。

（转下页）

Dear Sir

I refer to our conversation/your letter [*delete as appropriate*] of the [*insert date*] regarding my appointment as architect for the above project. I am happy to accept this commission.

I confirm that the terms and conditions which will apply to the project are those set out in the RIBA Conditions of Engagement for the Appointment of an Architect (CE/99), two copies of which are enclosed. They are completed in accordance with the terms we discussed with you.

Your attention is particularly drawn to the following:

1. The law of the contract will be the law of [*insert the relevant law, e.g. 'England and Wales' or 'Northern Ireland'*].

2. The time limit on the commencement of actions under this Agreement, as referred to in clause 7.2, will be [*insert the number of years, bearing in mind that the statutory period is 6 years for contracts under hand and 12 years for contracts executed as a deed*].

3. The limit of my liability, as referred to in clause 7.3, will be £[*insert the limit of your liability, taking into account your PI insurance and the likely construction cost*].

4. I maintain professional indemnity insurance cover of £[*insert amount*] for any one occurrence or series of occurrences arising out of one event.

5. Clause 9.5 deals with arbitration as the basic method of dispute resolution without prejudice to rights of adjudication. If we cannot agree on the name of an arbitrator, either of us may apply to the Royal Institute of British Architects for an appointment. [*If legal proceedings are preferred, substitute*:] We have agreed that clause 9.5 dealing with arbitration will not apply and that if a dispute arises, either of us may resort to legal proceedings.

6. The Effective Date of this Agreement when I will commence/commenced [*delete as appropriate*] services will be/is [*delete as appropriate*] [*insert date*].

[*continued*]

函件 3b 续表

Letter 3b continued

请核查一下各项服务和费用计算标准是否令你方满意。如果你方有什么疑问，请即刻与我方联系，以便给予解释。如果你方对各项条款都满意，请在此函的副本上签署协议条款，在表1的下方黄色标签所标明的区域内草签 CE/99 协议［*视情况，加上“并草签所有协议修正案”*］，并寄还给我方签字。之后，我方将寄给你方核证副本备案。

你忠诚的

协议

委托人希望建筑师进行与此项目有关的服务。雇佣条件为此信中列出的条款和随函附上的 RIBA 建筑师委任协议（CE/99）中规定的条款以及同时拟好的［*填入其他组成协议内容的文件名称*］。建筑师接受此协议。

委托人签字__________________ 日期__________________
建筑师签字__________________ 日期__________________

Please check that the services and basis of fee calculation is satisfactory. If you have any queries, please do not hesitate to ask me for clarification. If you are happy with all of this, please sign the Agreement clause on the copy of this letter, initial CE/99 at the foot of schedule 1［*if appropriate, add ', initial all amendments'*］where indicated by yellow sticker(s) and return all the documents to me for countersignature after which I will send you a certified copy for your records.

Yours faithfully

Agreement

The Client wishes to appoint the Architect to carry out Services in connection with the Project upon the terms and conditions set out in this letter and the attached copy of the RIBA Conditions of Engagement for the Appointment of an Architect（CE/99）as completed together with［*insert a description of any other papers which are to be part of the Agreement*］and the architect accepts.

Signed by the Client .... Date ....
Signed by the Architect .... Date ....

函件 3c

致委托人，说明 CE/99 委任条款

此函件仅在委托人为独立客户的情况下适用

Letter 3c

To client, setting out terms of appointment CE/99

This letter should only be used where the client is a consumer

尊敬的先生：

谨提及［填入日期］关于委任我方为以上项目建筑师的双方谈话/你方的来函［视情况取舍］。我方愿意接受这一委任，并深感荣幸。

现确认此项目将使用 RIBA 建筑师委任协议（CE/99）中规定的条款。现随函附上两份标准格式。这些条款是根据我方与你方的各自讨论和协商完成的。如果你方核查后对此感到满意，我方将不胜感激。

请特别注意以下几点：

1. 本合同遵守［填入相关法规，如“英格兰和威尔士法律”或“北爱尔兰法律”］。
2. 在第 7.2 条中提到，本协议规定的起诉时效为［填入年数，记住非盖印合约的法定年限为 6 年，契约合同的法定年限为 12 年］。
3. 在第 7.3 条中提到，我方的责任限额为［填入你方责任限额，把你方的专业责任保险（PI）及可能的建筑成本都计算在内］。
4. 我方对单一事件或单一事件引起的连续事件投保专业责任保险，投保金额为［填入金额］。
5. 第 9.4 条是关于快速解决争端机制的规定。这一机制引自《1996 年住房补贴，建造和维修法案》［当工程项目为北爱尔兰工程时，替换为《1997 年（北爱尔兰）建筑合同法令》］。［如果此法案或法令不适用，则加上：］尽管在此情况下裁决并非法定强制性程序，但我方确认你方希望采取此法律程序。如果双方不能就裁决人选达成一致，任何一方都可以向英国皇家建筑师协会申请指定裁决人。
6. 第 9.5 条规定了仲裁为解决争端的基本方法，这一规定不影响裁决权的使用。如果双方不能就仲裁人达成一致，任何一方都可以向英国皇家建筑协会申请指定仲裁人。［如果更希望通过法律程序来解决争端，则替换为：］双方同意关于仲裁的第 9.5 条不适用，如果发生争端，任何一方都可以诉诸法律程序。

（转下页）

Dear Sir

I refer to our conversation/your letter [*delete as appropriate*] of the [*insert date*] regarding my appointment as architect for the above project. I am happy to accept this commission.

I confirm that the terms and conditions which will apply to the project are those set out in the RIBA Conditions of Engagement for the Appointment of an Architect (CE/99), two copies of which are enclosed. They are completed in accordance with the terms we discussed and negotiated with you on an individual basis. I should be pleased if you would check that you are entirely happy with them.

Your attention is particularly drawn to the following:

1. The law of the contract will be the law of [*insert the relevant law, e. g. 'England and Wales' or 'Northern Ireland'*].

2. The time limit on the commencement of actions under this Agreement, as referred to in clause 7. 2, will be [*insert the number of years, bearing in mind that the statutory period is 6 years for contracts under hand and 12 years for contracts executed as a deed*].

3. The limit of my liability, as referred to in clause 7. 3, will be £ [*insert the limit of your liability, taking into account your PI insurance and the likely construction cost*].

4. I maintain professional indemnity insurance cover of £ [*insert amount*] for any one occurrence or series of occurrences arising out of one event.

5. Clause 9. 4 deals with a quick system of dispute resolution introduced by the Housing Grants, Construction and Regeneration Act 1996 [*substitute 'Construction Contracts (Northern Ireland) Order 1997' when dealing with a commission in Northern Ireland*]. [*If the Act or Order does not apply, add:*] Although adjudication is not obligatory under statute in this instance, I confirm that you wish to take advantage of the procedure. If we cannot agree on the name of an adjudicator, either of us may apply to the Royal Institute of British Architects for an appointment.

6. Clause 9. 5 deals with arbitration as the basic method of dispute resolution without prejudice to rights of adjudication. If we cannot agree on the name of an arbitrator, either of us may apply to the Royal Institute of British Architects for an appointment. [*If legal proceedings are preferred, substitute:*] We have agreed that clause 9. 5 dealing with arbitration will not apply and that if a dispute arises, either of us may resort to legal proceedings.

[*continued*]

函件 3c 续表

7. 本协议生效日期，即我方开始服务的日期，将为［视情况取舍］［填入日期］。

8. 表 1：项目描述栏中列出你方的要求。

9. 表 2：服务栏列出我方要履行的服务项目。我方预计必要的现场勘测次数为［填入次数］。其他活动栏中列出了我方可以提供的另外收费的服务项目。

10. 表 3：收费与费用栏列出了你方需要支付的服务费用及收费标准。

11. 表 4：其他委任栏中列出了我方认为你方必须雇佣的其他顾问，因为我方不具备这方面的专业经验。

如果你方对于各项条款都满意，请在此函的副本上签署协议条款，在表 1 的下方黄色标签所标明的区域内草签 CE/99 协议［视情况，加上“并草签所有协议修正案”］，并寄还给我方签字。之后，我方将寄给你方核证副本备案。

你忠诚的

协议

委托人希望建筑师进行与此项目有关的服务。雇佣条件为此信函中列出的条款和随函附上的 RIBA 建筑师委任协议（CE/99）中规定的条款以及同时拟好的［填入其他组成协议内容的文件名称］。建筑师接受此协议。

委托人签字＿＿＿＿＿＿＿＿＿＿ 日期＿＿＿＿＿＿＿＿＿＿

建筑师签字＿＿＿＿＿＿＿＿＿＿ 日期＿＿＿＿＿＿＿＿＿＿

［注：此合同须遵守《1999 年客户合同中的不公平条款规定》。应该注意的是，非双方各自商定的条款，如果不利于客户，都可能被认为是不公平的。正如此种情况下，如果各项条款为事先起草，则通常认为未经过双方各自协商。如果有关方面核查此事，你方须说明这些条款是经过双方各自协商而定的。］

Letter 3c continued

7. The Effective Date of this Agreement when I will commence/commenced [*delete as appropriate*] services will be/is [*delete as appropriate*] [*insert date*].

8. Schedule 1: the Project description records your requirements.

9. Schedule 2: the Services sets out the services I will perform. I expect that [*insert number*] visits to the Works will be necessary. Other activities list the services which I can provide at an additional charge.

10. Schedule 3: Fees and expenses shows the fees you will pay for the Services and the basis of charging.

11. Schedule 4: Other appointments shows a list of the other consultants I expect to be necessary for you to engage, because I do not have that particular expertise.

If you are happy with all of this, please sign the Agreement clause on the copy of this letter, initial CE/99 at the foot of schedule 1 [*if appropriate, add ', initial all amendments'*] where indicated by yellow sticker (s) and return all the documents to me for countersignature after which I will send you a certified copy for your records.

Yours faithfully

Agreement

The Client wishes to appoint the Architect to carry out Services in connection with the Project upon the terms and conditions set out in this letter and the attached copy of the RIBA Conditions of Engagement for the Appointment of an Architect (CE/99) as completed together with [*insert a description of any other papers which are to be part of the Agreement*] and the architect accepts.

Signed by the Client .... Date ....
Signed by the Architect .... Date ....

[*Note: This contract is subject to the Unfair Terms in Consumer Contracts Regulations 1999. It should be noted that terms which are not individually negotiated may be regarded as unfair if unbalanced to the detriment of the consumer. If the terms have been drafted in advance, such as in this instance, they will always be considered as not having been individually negotiated and, if challenged, it is for you to show that they were.*]

函件 3d

致委托人，说明 SW/99 委任条款

当委托人为独立客户时，此函件不适用

Letter 3d

To client, setting out terms of appointment SW/99

This letter should not be used where the client is a consumer

尊敬的先生：

谨提及［填入日期］关于委任我方为以上项目建筑师的双方谈话/你方来函［视情况取舍］。我方愿意接受这一委任，并深感荣幸。

现确认此项目使用 RIBA 小型建筑建筑师委任协议（SW/99）中规定的条款。现随函附上两份标准格式。这些条款是根据我们双方讨论商定的条款拟好的。

请特别注意以下几点：

1. 本合同遵守［填入相关法规，如“英格兰和威尔士法律”或“北爱尔兰法律”］。
2. 在第 7.2 条中提到，本协议规定的起诉时效为［填入年数，记住非盖印合约的法定年限为 6 年，契约合同的法定年限为 12 年］。
3. 在第 7.3 条中提到，我方的责任限额为［填入你方责任限额，把你方的专业责任保险（PI）及可能的建筑成本都计算在内］。
4. 我方对单一事件或单一事件引起的连续事件投保专业责任保险，投保金额为［填入金额］。
5. 我方将履行服务清单中列出的服务项目。我方预计必要的现场勘测次数为［填入次数］。其他活动栏中列出了我方可以提供的另外收费的服务项目。
6. 费用收取将按百分比/总金额/时间［视情况取舍］来计算。

［然后加上：］

收费的百分比为：总建筑成本的［填入数字］。

［或者：］

收费总金额为［填入金额］。

（转下页）

Dear Sir

I refer to our conversation/your letter [*delete as appropriate*] of the [*insert date*] regarding my appointment as architect for the above project. I am happy to accept this commission.

I confirm that the terms and conditions which will apply to the project are those set out in the RIBA Conditions of Appointment for Small Works (SW/99), two copies of which are enclosed. They are completed in accordance with the terms we discussed with you.

Your attention is particularly drawn to the following:

1. The law of the contract will be the law of [*insert the relevant law, e.g. 'England and Wales' or 'Northern Ireland'*].

2. The time limit on the commencement of actions under this Agreement will be [*insert the number of years, bearing in mind that the statutory period is 6 years for contracts under hand and 12 years for contracts executed as a deed*].

3. The limit of my liability will be £ [*insert the limit of your liability, taking into account your PI insurance and the likely construction cost*].

4. I maintain professional indemnity insurance cover of £ [*insert amount*] for any one Occurrence or series of occurrences arising out of one event.

5. I will carry out the services so indicated on the Schedule of Services. I expect that [*insert number*] visits to the Works will be necessary. Other activities list the services which I can provide at an additional charge.

6. Fees will be charged on a percentage/lump sum/time [*delete as appropriate*] basis.

[*Then add either*:]

The percentage will be [*insert figure*] % of the total Construction Cost.

[*Or*:]

The lump sum will be £ [*insert amount*].

[*continued*]

函件 3d 续表

[然后加上:]

按时间收取费用标准为[列出该项目涉及的各类员工的小时报酬细节]。

7. 与此项委任相关的所有费用将按照我方通常的收费标准收取。

8. 我方的收费及支出费用都需交纳增值税(VAT)[或:]我方尚未注册交纳增值税,但如果在委任期间,我方进行了注册,那么我方的收费和支出费用都需交纳增值税。

9. 第39条规定了仲裁为解决争端的基本方法,这一规定不影响裁决权的使用。如果双方不能就仲裁人选达成一致,任何一方都可以向英国皇家建筑师协会申请指定仲裁人。[如果更希望通过法律程序来解决争端,则替换为:]双方同意关于仲裁的第39条不适用,如果发生争端,任何一方都可以诉诸法律程序。

10. 本协议生效日期,即我方开始服务的日期,将为/[视情况取舍] [填入日期]。

请核查一下各项服务和费用计算标准是否满意。如果你方有什么疑问,请即刻与我方联系,以便给予解释。如果你方对于各项条款都满意,请在此函的副本上签署协议条款,在表1的下方黄色标签所标明的区域内草签SW/99协议[视情况,加上"并草签所有协议修正案"],寄还给我方签字。之后,我方将寄给你方核证副本备案。

你忠诚的

协议

委托人希望建筑师进行与此项目有关的服务。雇佣条件为此信函中列出的条款和随函附上的RIBA建筑师委任协议(CE/99)中规定的条款以及同时拟好的[填入其他组成协议内容的文件名称]。建筑师接受此协议。

委托人签字________________ 日期________________

建筑师签字________________ 日期________________

Letter 3d continued

[*Then add*:]

The time charge will be [*set out details of the hourly rates for all categories of staff who will be working on the project*].

7. All expenses of whatever kind incurred in connection with this appointment will be charged on the fees at my usual rates.

8. [*Either*:] VAT is chargeable on all my fees and expenses. [*Or*:] I am not registered for VAT, but VAT will be charged on fees and expenses if, during the appointment, I become registered.

9. Clause 39 deals with arbitration as the basic method of dispute resolution without prejudice to rights of adjudication. If we cannot agree on the name of an arbitrator, either of us may apply to the Royal Institute of British Architects for an appointment. [*If legal proceedings are preferred, substitute*:] We have agreed that clause 39 dealing with arbitration will not apply and that if a dispute arises, either of us may resort to legal proceedings.

10. The date of this Agreement when I will commence/commenced [*delete as appropriate*] services will be/is [*delete as appropriate*] [*insert date*].

Please check that the services and basis of fee calculation is satisfactory. If you have any queries, please do not hesitate to ask me for clarification. If you are happy with all of this, please sign the Agreement clause on the copy of this letter, initial SW/99 at the foot of the Schedule of Services and the Conditions [*if appropriate, add '*, *initial all amendments'*] where indicated by yellow sticker (s) and return all the documents to me for countersignature after which I will send you a certified copy for your records.

Yours faithfully

Agreement

The Client wishes to appoint the Architect to carry out Services in connection with the Project upon the terms and conditions set out in this letter and the attached copy of the RIBA Conditions of Engagement for the Appointment of an Architect (CE/99) as completed together with [*insert a description of any other papers which are to be part of the Agreement*] and the architect accepts.

Signed by the Client .... Date ....

Signed by the Architect .... Date ....

函件 3e

致委托人，说明 SW/99 委任条款

此函件仅在委托人为独立客户的情况下适用

Letter 3e

To client, setting out terms of appointment SW/99

This letter should only be used where the client is a consumer

尊敬的先生：

谨提及［填入日期］关于委任我方为以上项目建筑师的双方谈话/你方来函［视情况取舍］。我方愿意接受这一委任，并深感荣幸。

现确认适用于此项目的条款为 RIBA 小型建筑建筑师委任协议（SW/99）中规定的条款。现随函附上两份标准格式。这些条款是根据我方与你方各自商谈讨论的内容完成的。如果你方核查后对此感到满意，我方将不胜感激。

请你方特别注意以下各项：

1. 本合同遵守［填入相关法规，如“英格兰和威尔士法律”或“北爱尔兰法律”］。
2. 在第 7.2 条中提到，本协议规定的起诉时效为［填入年数，记住非盖印合约的法定年限为 6 年，契约合同的法定年限为 12 年］。
3. 在第 7.3 条中提到，我方的责任限额为［填入你方责任限额，把你方的专业责任保险（PI）及可能的建筑成本都计算在内］。
4. 我方认为应针对单一事件或单一事件引起的连续事件投保专业责任保险，投保金额为［填入金额］。
5. 我方将履行服务栏中列出的服务项目。我方预计的必要的现场勘测次数为［填入次数］。其他活动栏中列出了我方可以提供的另外收费的服务项目。
6. 费用收取将按百分比/总金额/时间［视情况取舍］来计算。

［然后加上：］

收费的百分比为总建筑成本的［填入数字］%。

（转下页）

Dear Sir

I refer to our conversation/your letter [*delete as appropriate*] of the [*insert date*] regarding my appointment as architect for the above project. I am happy to accept this commission.

I confirm that the terms and conditions which will apply to the project are those set out in the RIBA Conditions of Appointment for Small Works (SW/99), two copies of which are enclosed. They are completed in accordance with the terms we discussed and negotiated with you on an individual basis. I should be pleased if you would check that you are entirely happy with them.

Your attention is particularly drawn to the following:

1. The law of the contract will be the law of [*insert the relevant law, e.g. 'England and Wales' or 'Northern I* reland'].

2. The time limit on the commencement of actions under this Agreement will be [*insert the number of years, bearing in mind that the statutory period is 6 years for contracts under hand and 12 years for contracts executed as a deed*].

3. The limit of my liability will be £ [*insert the limit of your liability, taking into account your PI insurance and the likely construction cost*].

4. I maintain professional indemnity insurance cover of £ [*insert amount*] for any one occurrence or series of occurrences arising out of one event.

5. I will carry out the services so indicated on the Schedule of Services. I expect that [*insert number*] visits to the Works will be necessary. Other activities list the services which I can provide at an additional charge.

6. Fees will be charged on a percentage/lump sum/time [*delete as appropriate*] basis.

[*Then add either*:]

The percentage will be [*insert figure*] % of the total Construction Cost.

[*continued*]

函件 3e 续表

［或者：］

收费总金额为［填入金额］。

［然后加上：］

按时间收取费用标准为：［列出该项目涉及的各类员工的小时报酬细节］

7. 与此项委任相关的所有费用将按照我方通常的收费标准收取。

8. 我方的收费及支出费用都需交纳增值税（VAT）［或：］我方尚未注册交纳增值税，但如果在委任期间，我方进行了注册，我方的收费和支出费用都需交纳增值税。

9. 第 37 条是关于快速解决争端机制的规定。这一机制引自《1996 年住房补贴、建造和维修法案》［当工程项目为北爱尔兰工程时，替换为《1997 年（北爱尔兰）建筑合同法令》］。［如果此法案或法令不适用，则加上：］尽管在此情况下裁决并非法定强制性程序，但我方确认你方希望采取此法律程序。如果双方不能就裁决人选达成一致，任何一方都可以向英国皇家建筑协会申请指定裁决人。

10. 第 39 条规定了仲裁为解决争端的基本方法，这一规定并不影响裁决权的使用。如果双方不能就仲裁人选达成一致，任何一方都可以向英国皇家建筑师协会申请指定仲裁人。［如果更希望通过法律程序来解决争端，则替换为：］双方同意关于仲裁的第 39 条不适用，如果发生争端，任何一方都可以诉诸法律程序。

（转下页）

Letter 3e continued

[*Or*:]

The lump sum will be £ [*insert amount*].

[*Then add*:]

The time charge will be [*set out details of the hourly rates for all categories of staff who will be working on the project*].

7. All expenses of whatever kind incurred in connection with this appointment will be charged on the fees at my usual rates.

8. [*Either*:] VAT is chargeable on all my fees and expenses. [*Or*:] I am not registered for VAT, but VAT will be charged on fees and expenses if, during the appointment, I become registered.

9. Clause 37 deals with a quick system of dispute resolution introduced by the Housing Grants, Construction and Regeneration Act 1996 [*substitute 'Construction Contracts (Northern Ireland) Order 1997' when dealing with a commission in Northern Ireland*]. [*If the Act or Order does not apply, add*:] Although adjudication is not obligatory under statute in this instance, I confirm that you wish to take advantage of the procedure. If we cannot agree on the name of an adjudicator, either of us may apply to the Royal Institute of British Architects for an appointment.

10. Clause 39 deals with arbitration as the basic method of dispute resolution without prejudice to rights of adjudication. If we cannot agree on the name of an arbitrator, either of us may apply to the Royal Institute of British Architects for an appointment. [*If legal proceedings are preferred, substitute*:] We have agreed that clause 39 dealing with arbitration will not apply and that if a dispute arises, either of us may resort to legal proceedings.

[*continued*]

函件 3e 续表

11. 本协议生效日期，即我方开始服务的日期，将为［视情况取舍］［填入日期］。

请核查一下各项服务和费用计算标准是否满意。如果你方有什么疑问，请即刻与我方联系，以便给予解释。如果你方对于各项条款都满意，请在此函的副本上签署协议条款，在表 1 的下方黄色标签所标明的区域内草签 SW/99 协议［视情况，加上“并草签所有协议修正案”］，并寄还给我方签字。之后，我方将寄给你方核证副本备案。

协议

委托人希望建筑师进行与此项目有关的服务。雇佣条件为此信函中列出的条款和随函附上的 RIBA 建筑师委任协议（CE/99）中规定的条款以及同时拟好的［填入其他组成协议内容的文件名称］。建筑师接受此协议。

委托人签字＿＿＿＿＿＿＿＿　　日期＿＿＿＿＿＿＿＿
建筑师签字＿＿＿＿＿＿＿＿　　日期＿＿＿＿＿＿＿＿

［注：此合同须遵守《1999 年客户合同中的不公平条款规定》。应该注意的是，非双方各自商定的条款，如果不利于客户，都可能被认为是不公平的。正如此种情况下，如果各项条款为事先起草，则通常认为未经过双方各自协商。如果有关方面核查此事，你方须说明这些条款是经过双方各自协商而定的。］

你忠诚的

Letter 3e continued

11. The date of this Agreement when I will commence/commenced [*delete as appropriate*] services will be/is [*delete as appropriate*] [*insert date*].

Please check that the services and basis of fee calculation is satisfactory. If you have any queries, please do not hesitate to ask me for clarification. If you are happy with all of this, please sign the Agreement clause on the copy of this letter, initial SW/99 at the foot of the Schedule of Services and the Conditions [*if appropriate, add ', initial all amendments'*] where indicated by yellow sticker (s) and return all the documents to me for countersignature after which I will send you a certified copy for your records.

Yours faithfully

Agreement

The Client wishes to appoint the Architect to carry out Services in connection with the Project upon the terms and conditions set out in this letter and the attached copy of the RIBA Conditions of Engagement for the Appointment of an Architect (CE/99) as completed together with [*insert a description of any other papers which are to be part of the Agreement*] and the architect accepts.

Signed by the Client .... Date ....

Signed by the Architect .... Date ....

[*Note: This contract is subject to the Unfair Terms in Consumer Contracts Regulations 1999. It should be noted that terms which are not individually negotiated may be regarded as unfair if unbalanced to the detriment of the consumer. If the terms have been drafted in advance, such as in this instance, they will always be considered as not having been individually negotiated and, if challenged, it is for you to show that they were.*]

函件 4

致委托人，要求预先付款

Letter 4

To client, requesting payment in advance

尊敬的先生：

你方于［填入日期］的关于指定我方履行以上项目相关服务的来函收悉，深表谢意。

如果能去你处/在我处［视情况取舍］与你方面谈详细要求及委任条款，我方将深感荣幸。希望你方在双方签订协议之时，预付预计总费用的［填入百分比］[1]。

随函附上一份 RIBA 建筑师委任协议标准格式（SFA/99）［当委托项目为中型工程时应为“建筑师委任协议（CE/99）”，小型工程时应为“小型建筑建筑师委任协议（SW/99）”］。请你方在审查后，打电话联系我方，以便商定一个合适的面谈时间。

你忠诚的

［1 如果填入具体数目而不是百分比，可能会避免争端。］

Dear Sir

Thank you for your letter of the [*insert date*] instructing me to carry out architectural services in connection with the above project.

I should be pleased to visit you/see you at this office [*delete as appropriate*] to discuss your detailed requirements and my terms of appointment. I would ask for a payment on account of [*insert percentage*] of the estimated total fees[1] at the time of signing the agreement between us.

A copy of the RIBA Standard Form of Agreement for the Appointment of an Architect (SFA/99) [*substitute* '*Conditions of Engagement* (*CE/99*)' *when dealing with medium sized commissions and* '*Conditions of Appointment for Small Works* (*SW/99*)' *when dealing with small commissions*] is enclosed. After you have had the opportunity to examine it, perhaps you will telephone me to arrange a convenient date and time for our meeting.

Yours faithfully

[[1] *You may find it prevents dispute if you insert an actual sum instead of a percentage.*]

函件 5

关于该工程雇佣的其他建筑师，致委托人

Letter 5

To client, regarding other architects engaged on the work

尊敬的先生：

为了遵守 RIBA 的《职业操守规范》，我方有责任询问你方，在我方之前是否为该项目雇佣过其他建筑师。

无论在任何时候，如果该项目涉及到另一位建筑师，请你方将他/她［视情况取舍］的姓名和地址告知我方，以便我方将“现受你方委任为建筑师”的事实通知他/她［视情况取舍］。

你忠诚的

Dear Sir

In order to comply with the RIBA Code of Professional Conduct, I am required to make reasonable enquiries to discover if you have previously engaged any other architect on this project.

If another architect has been involved at any time, perhaps you will let me have his/her [*delete as appropriate*] name and address so that I can inform him/her [*delete as appropriate*] that I am now acting for you.

Yours faithfully

函件 6

如果委托人要求建筑师签署担保书，致委托人

Letter 6

To client, if architect asked to sign a warranty

尊敬的先生：

你方于［填入日期］的来函及其附件担保书收悉，深表谢意。

在委任我方为建筑师的委任条款中，没有规定我方必须签署担保协议。相信你方知道签署这种担保书会大大增加我方的潜在责任风险。

不过，我方承认担保协议的重要性，尤其是涉及到销售或全额维修租赁的情况时。因此，我方准备签署一份条款合理的担保协议，但你方必须为此额外支付一笔费用。随函附上一份简单的担保书标准格式。很抱歉其中各项条款不可协商，请告知你方是否接受这些条款，我方将对此收取［填入要求的费用金额，数目不可过小］。

你忠诚的

Dear Sir

Thank you for your letter of the [*insert date*] with which you enclosed a form of warranty for signature.

There is nothing in the terms of engagement governing my appointment which obliges me to enter into a warranty agreement. I am sure you understand that if I was to execute such a warranty, the scope of my potential liability would be substantially increased.

However, I appreciate that a warranty agreement is a valuable commodity, particularly where sale or full repairing lease is concerned. Therefore, I am prepared to enter into a warranty in appropriate terms on payment of an additional fee as consideration. A copy of a simple standard form of warranty is enclosed and I should be pleased if you would let me know if its terms are acceptable. I am afraid that they are not negotiable in this instance. My fee would be [*insert fee required which should be more than nominal*].

Yours faithfully

函件 7

关于以前的雇佣关系，致其他建筑师

Letter 7

To other architect, regarding former engagement

尊敬的先生：

谨告知我方已被［填入委托人姓名］确定为上述项目的建筑师。

获悉你方曾经被雇佣担任此项目的建筑师，请视此函为根据 RIBA 的《职业操守规范》第三条第 3.8 款发出的正式通知。

若你方有任何意见，希望能告知我方，我方将深感谢意。

你忠诚的

Dear Sir

I have been approached by [*insert name of client*] to undertake the above project.

I understand that you were engaged on this project at one time and you should take this letter as notice in accordance with principle 3, rule 3.8 of the RIBA Code of Professional Conduct.

If you have any comments, I should be pleased to receive them.

Yours faithfully

函件 8

如果前任建筑师告知有关问题，致委托人

Letter 8

To client, if former architect notifies some problem

尊敬的先生：

从［填入前任建筑师的姓名］处得知［填入问题性质］。

这显然不属于我方直接的职责范围，但是你方可以决定是否希望由我方全权负责处理所有的情况。

［在某些情况下，添加以下内容方为明智之举：］

但是，在这种情况下，且不评论在之前的雇佣关系中双方各自的责任，如果需要我方承担此项委托，我方要求你方支付的费用为［填入金额］。

你忠诚的

Dear Sir

I have been informed by [*insert name of former architect*] that [*insert nature of problem*].

Clearly this is not my direct concern but, entirely at your discretion, you may wish to appraise me of the full circumstances.

[*In some circumstances it may be wise to add the following*:]

In this instance, however, and without passing any comment on the respective liabilities under the previous engagement, I will require a payment on account of [*insert amount*] before I undertake this commission.

Yours faithfully

第二章　基本设计要点

在这一阶段，委托人将确认关键性要求和限制条件，并决定委任顾问事项以及具备设计资格的分包商人选。

在这一阶段或者以后的阶段中，有时会出现这种问题：委托人不希望因为要获得所有必要的批准而拖延进度，而是希望工程继续进行。从委托人的观点来看，这样可以节省时间和资金。但是如果没能获得相应的批准，建筑师则希望对窝工的工作索取费用赔偿。

函件 9

关于工地勘测，致委托人

Letter 9

To client, regarding site survey

尊敬的先生：

根据你方［填入日期］的电话指示，我方将进行工地勘测，并完成面积测量及水平高度测量。

此项工作不属于我方的正常服务范围，为此，我方将按照雇佣条款中的规定，按照工作时间额外收取费用。

你忠诚的

Dear Sir

I confirm your instructions by telephone on the [*insert date*] to carry out a site survey complete with measurements and levels.

This work is additional to my normal services, and extra fees and expenses will be chargeable on a time basis as indicated in the terms of engagement.

Yours faithfully

函件 10

进行工地勘测之前，致委托人

Letter 10

To client, before carrying out a site survey

尊敬的先生：

在我方对上述工地进行勘测以前，必须询问你方的土地所有权范围。这种问题经常不够明确，但可能对工程项目产生重大的影响。

随函附上一份平面概图，主要包括以下方面：墙、道路、邻近建筑等。请你方在图中场地周围用红线标出包括土地边界在内的所有权限的准确范围，并请尽快寄回图纸。如果可以的话，请标出尺寸。

如果随你方地契附上的平面图已经明确说明了所有权限问题，请告知我方可以在何处审查它。如果你方对所有权限有任何疑问，建议请你方的律师标出边界。

你忠诚的

Dear Sir

Before I carry out a survey of the above site, it is essential that I satisfy myself regarding the limits of your ownership. Such matters are frequently obscure, but they may have a crucial effect on the project.

A sketch plan is enclosed on which are noted the principal features: walls, roads, adjacent buildings, etc. I should be pleased if you would indicate the precise extent of your ownership, including the ownership of boundaries, by drawing a red line around the site and returning the sketch to me as soon as possible. Please show dimensions if available.

If the plan attached to your title deeds shows all this information clearly, please let me know where I can examine it. If you are in any doubt about your ownership, I advise you to ask your solicitor to indicate the boundaries for you.

Yours faithfully

函件 11

如果委托人请求建筑师协助进行边界谈判，致委托人

Letter 11

To client, if requested to help in boundary negotiations

尊敬的先生：

你方于［填入日期］的来函收悉，深表谢意。

我方很高兴能协助你方与相邻业主谈判以界定边界。如果由你方律师负责谈判，而我方到场给予建议，将对你方非常有利。

此项活动为额外服务［视情况，填入"雇佣条款规定的服务"］，我方将按每小时［填入金额］另外收取费用。希望你方能同意。

请通知你方律师直接电话联系我方，以商定面谈事宜。

你忠诚的

Dear Sir

Thank you for your letter of the [*insert date*].

I will be happy to assist in negotiations with adjoining owners in order to fix site boundaries. It will be in your own best interests if your own solicitor takes charge of the negotiations and I am present to advise.

This is an additional service [*insert, if appropriate*: '*to the services already agreed in the terms of engagement*'] and an additional fee at the rate of [*insert amount*] per hour is chargeable and I should be pleased to have your agreement.

Please ask your solicitor to telephone me directly to arrange a meeting.

Yours faithfully

函件 12

如果在勘测中遇到问题，致委托人

Letter 12

To client, if problem encountered during survey

尊敬的先生：

谨提及我们于［填入日期］的电话谈话。

在对上述场地进行勘测的过程中，我方发现如下问题［填入对问题的简洁明了的描述］。

鉴于此问题与该项目的密切关系及其影响，我方希望尽快在工地与你方面谈。［视情况，加上：］我方将邀请［填入顾问或承包商的名称］到场，以便迅速做出决定。

明日前后，我方将安排好必要事宜，并电话通知你方。

你忠诚的

Dear Sir

I refer to our telephone conversation of the [*insert date*].

During my survey of the above property, it was discovered that [*insert a clear and concise description of the problem*].

In view of the implications and the effect upon the project, it is desirable that I meet you on site urgently. [*Add, if appropriate:*] I will invite [*insert name of consultant or contractor*] to be present to facilitate an immediate decision.

I will telephone you during the next day or so when I have made the necessary arrangements.

Yours faithfully

函件 13

致地质专家，询问土质勘测情况

Letter 13

To geotechnical specialists, enquiring about soil survey

尊敬的先生：

我方是负责上述项目的建筑师。我方的委托人为［填入委托人姓名］。

我方受委托人的授权，询问你方进行工地土质勘测的报价。

此项目包括［填入对项目规模和特征的简要描述］。设计规划定于［填入重要日期］进行，因此，土质勘测必须在［填入日期］之前完成，并同时将勘测报告交于我方。

请告知你方将进行勘测的细节内容和费用，以及相关的卫生和安全手续细节。你方占用工地及进行所有必要的工作时，需要考虑到公众安全和财产保护事项。同时你方须进行投保，以保证我方的委托人对于因你方的工地占用以及工地工作引起的索赔事件不负任何责任。随函附上编号为［填入编号］的工地平面图作为参考。

你忠诚的

Dear Sir

I act as architect for the above development. My client is [*insert name*].

My client has authorised me to seek quotations for the carrying out of a ground investigation on the site.

The development will consist of [*insert brief description of the size and character of the project*]. The design programme is [*insert key dates*] and, therefore, the ground investigation must be complete and in my hands by [*insert date*].

Please let me have details of the investigations you would carry out and their cost and details of your health and safety procedures. You will be expected to include for all necessary work and attendance, protection of the public and property and take out insurances and indemnify my client against any claims arising from your occupation of or work on the site. Site plan number [*insert number*] is enclosed for your information.

Yours faithfully

函件 14

致煤炭管理部门，询问初步信息

Letter 14

To Coal Authority, requesting preliminary information

尊敬的先生：

我方是负责上述项目的建筑师。我方的委托人为［填入委托人姓名］。

随函附上两份编号为［填入编号］的工地平面图和一份总平面图。倘若你方能告知该工地是否有可能受到你方过去、现在或未来计划的作业影响，我方将不胜感激。该项目预期为［描述建筑物的总体特征，包括楼层数在内］。

随函寄去付与你方的费用，共计［填入金额］。

根据《1991 年煤矿津贴法案》第 34(2)(a)条款，此函即代表建筑物所有者向你方发出的通知。

你忠诚的

抄送：委托人

Dear Sir

I act as architect for the above development. My client is [*insert name*].

Two copies of the site plan number [*insert number*] and a general location plan of the area are enclosed. I should be pleased if you would inform me if this site is likely to be affected by your operations, past, present or projected. The proposed development is expected to be [*describe general character of the building including number of storeys*].

I enclose your fee in the sum of [*insert amount*].

Please consider this as notice on behalf of the building owner in accordance with the notice requirements of the Coal Mining Subsidence Act 1991, section 34(2)(a).

Yours faithfully

Copy: Client

函件 15

致电话业务供应商，询问初步信息

Letter 15

To telephone service supplier, requesting preliminary information

尊敬的先生：

我方是负责上述项目的建筑师。我方的委托人为［填入委托人姓名］。

随函附上两份［填入编号］工地平面图。如果你方能予以审查，并告知我方以下信息，我方将不胜感激。

［视情况删去下列任何条目］

1. 工地内及邻近的通信电缆及设施的位置、地下深度或地上高度。电线杆或其他设施的位置。
2. 任何可能需要我方委托人支付的服务费用。
3. 电话设施的位置及电缆入口管道的要求。
4. 任何其他特殊的要求。

你忠诚的

Dear Sir

I act as architect for the above development. My client is [*insert name*].

Two copies of the site plan number [*insert number*] are enclosed. I should be pleased if you would examine it and let me have the following information:

[*Delete as appropriate from the following*:]

1. The position and depth below ground or height above ground of all telephone cables and services and the position of all poles or other equipment on or adjacent to the site.
2. Any contribution to the cost of service which may be required from my client.
3. Requirements with regard to service positions and inlet ducts.
4. Any other special requirements.

Yours faithfully

函件 16

致电力供应商，询问初步信息

Letter 16

To electricity supplier, requesting preliminary information

尊敬的先生：

我方是负责上述项目的建筑师。我方的委托人为［填入委托人姓名］。

随函附上两份［填入编号］工地平面图。如果你方能予以审查，并告知我方以下信息，我方将不胜感激。

［视情况删去下列任何条目］

1. 工地内或邻近的电力干线及已知电力设施的位置、大小、深度或高度。
2. 需要我方委托人支付的费用金额。
3. 测量及电线入口管道的要求。
4. 任何其他特殊要求。

你忠诚的

Dear Sir

I act as architect for the above development. My client is [*insert name*].

Two copies of the site plan number [*insert number*] are enclosed. I should be pleased if you would examine it and let me have the following information:

[*Delete as appropriate from the following*:]

1. The position, size and depth of all electricity mains and known services on or adjacent to the site.
2. The amount of contribution to the cost which will be required from my client.
3. Requirements with regard to metering and inlet ducts.
4. Any other special requirements.

Yours faithfully

函件 17
致燃气供应商，询问初步信息

Letter 17
To gas supplier, requesting preliminary information

尊敬的先生：

我方是负责上述项目的建筑师。我方的委托人为［填入委托人姓名］。

随函附上两份［填入编号］工地平面图。如果你方能予以审查，并告知我方以下信息，我方将不胜感激。

［视情况删去下列任何条目］

1. 工地内或邻近的煤气管道及已知煤气设施的位置、大小、深度或高度。
2. 需要我方委托人支付的费用金额。
3. 测量及入口管道的要求。
4. 任何其他特殊要求。

你忠诚的

Dear Sir

I act as architect for the above development. My client is [*insert name*].

Two copies of the site plan number [*insert number*] are enclosed. I should be pleased if you would examine it and let me have the following information:

[*Delete as appropriate from the following*:]

1. The position, size and depth of all gas mains and known services on or adjacent to the site.
2. The amount of contribution to the cost of supply which will be required from my client.
3. Requirements with regard to metering and inlet ducts.
4. Any other special requirements.

Yours faithfully

函件 18

致自来水供应商，询问初步信息

Letter 18

To water supplier, requesting preliminary information

尊敬的先生：

我方是负责上述项目的建筑师。我方的委托人为［填入委托人姓名］。

随函附上两份［填入编号］工地平面图。如果你方能予以审查，并告知我方以下信息，我方将不胜感激。

［视情况删去下列任何条目］

1. 工地内或邻近的水力管道及已知水力设施的位置、大小、深度或高度。
2. 未来自来水供应的预计水压。
3. 需要我方委托人支付的费用金额。
4. 是否需要水力测量。
5. 水力储备要求。
6. 地方水力规定。

你忠诚的

Dear Sir

I act as architect for the above development. My client is [*insert name*].

Two copies of site plan number [*insert number*] are enclosed. I should be pleased if you would examine it and let me have the following information:

[*Delete as appropriate from the following*:]

1. The position, size and depth of all water mains and known services on or adjacent to the site.
2. The anticipated water pressure in the proposed supply.
3. The amount of contribution to the cost which will be required from my client.
4. Whether water metering is required.
5. Water storage requirements.
6. Local water regulations.

Yours faithfully

函件 19

致排污管理部门，询问初步信息

Letter 19

To drainage authority, requesting preliminary information

尊敬的先生：

我方是负责上述项目的建筑师。我方的委托人为［填入委托人姓名］。

随函附上两份工地平面图［填入号码］。如果你方能予以审查，并告知我方以下信息，我方将不胜感激。

［视情况删去下列任何条目］

1. 工地内或邻近的地面排水及下水道的位置、大小、顺序和流向。
2. 要求的排污网络系统以及排污部门是否打算实施该工程建设。
3. 该工地的污水及地表水排出系统。
4. 任何具体的有关该地区排水的要求。
5. 技术规范要求。
6. 任何不正常的工地状况，如，下水道污水泛滥或外溢。

你忠诚的

Dear Sir

I act as architect for the above development. My client is [*insert name*].

Two copies of site plan number [*insert number*] are enclosed. I should be pleased if you would examine it and let me have the following information:

[*Delete as appropriate from the following*:]

1. Positions, sizes, inverts and flow of all surface water and foul sewers on or adjacent to the site.
2. System of drainage connection required and whether the authority wishes to carry out this part of the work.
3. System of foul and surface water drainage from the site.
4. Any particular requirements with regard to drainage in this area.
5. Specification requirements.
6. Any unusual site conditions such as flooding or surcharging of sewers.

Yours faithfully

函件 20

致公路管理部门，询问初步信息

Letter 20

To highway authority, requesting preliminary information

尊敬的先生：

我方是负责上述项目的建筑师。我方的委托人为［填入委托人姓名］。

随函附上两份［填入编号］工地平面图。如果你方能予以审查，并告知我方以下信息，我方将不胜感激。

［视情况删去下列任何条目］

1. 任何可能影响到该工地或在工地附近计划建设的汽车高速公路的位置。
2. 任何可能影响到该工地或在工地附近计划建设的公路的位置。
3. 任何可能影响到该工地现有或计划建设的建筑物或修缮改造道路的位置。
4. 该工地可接受的出入口位置。
5. 该项目有望于［填入日期］之前完成。如能收到你方到完工时仍有效的公路技术说明，我方将不胜感激。
6. 可能影响到此项目的任何活动的细节。

你忠诚的

Dear Sir

I act as architect for the above development. My client is [*insert name*].

Two copies of the site plan number [*insert number*] are enclosed. I should be grateful if you would examine it and let me have the following information:

[*Delete as appropriate from the following*:]

1. The position of any future motorway schemes which might affect the site or which are proposed in the vicinity of the site.
2. The position of any future highway schemes which might affect the site or which are proposed in the vicinity of the site.
3. The position of any building or improvement lines, existing or proposed, which might affect the site.
4. Acceptable positions for ingress to and egress from the site.
5. It is hoped to complete the development by [*insert date*] and I should be pleased to receive a copy of your highway specification which will be current at that date.
6. Details of any further operations which might affect this development.

Yours faithfully

函件 21

致电力供应商，询问初步信息

Letter 21

To power supplier, requesting preliminary information

尊敬的先生：

我方是负责上述项目的建筑师。我方的委托人为［填入委托人姓名］。

随函附上两份［填入编号］工地平面图。如果你方能予以审查，并告知是否在工地内或工地附近配备有或将来有计划配备可能会影响到工地的任何设备，对此我方将不胜感激。

你忠诚的

Dear Sir

I act as architect for the above development. My client is [*insert name*].

Two copies of the site plan number [*insert number*] are enclosed. I should be pleased if you would examine it and let me know if you have any equipment on or adjacent to the site or any proposals for the future which might affect the site.

Yours faithfully

函件 22

致委托人，寻求信息

Letter 22

To client, seeking information

尊敬的先生：

根据我们［填入日期］的会谈，我方现确认你方将自行进行调查，请在［填入日期］之前为我方提供下列信息，以便我方能完成对此项目的可行性研究。

1. 场地/地产［视情况取舍］的所有权类别。
2. 限制性契约的细节。
3. 工地或与工地相关的地役权细节，如，道路使用权、采光权、架设权等。
4. 现有的工地/地产［视情况取舍］的图纸。
5. 已经获得的工地/地产［视情况取舍］规划许可或其他许可。

我方期望能在［填入日期］之前完成可行性报告

你忠诚的

Dear Sir

Following our meeting on the [*insert date*], I confirm that you will carry out your own investigations to provide me with the following information by [*insert date*], so that I can complete my feasibility studies into this project:

1. The type of ownership of the site/property [*delete as appropriate*].
2. Details of restrictive covenants.
3. Details of easements over the site or in connection with the site, for example: rights of way, rights of light, rights of support, etc.
4. Existing drawings of the site/property [*delete as appropriate*].
5. Planning or other approvals already obtained for this site/property [*delete as appropriate*].

I anticipate completing my feasibility report by [*insert date*].

Yours faithfully

函件 23

致委托人，随函附有可行性报告

Letter 23

To client, enclosing the feasibility report

尊敬的先生：

随函附上以上项目的可行性报告［填入份数］份。谨确认我方将于［填入日期］去你处讨论此报告内容，并听取你方的指示。

［视情况，加上：］

请特别注意［填入最重要的事项］。

你忠诚的

Dear Sir

I enclose [*insert number*] copies of my feasibility report for the above project. I confirm that I will visit you on the [*insert date*] at [*insert time*] to discuss the contents and to take your instructions.

[*If appropriate*, *add*:]

May I draw your attention particularly to [*insert whatever matter is most important*].

Yours faithfully

函件 24

关于早期顾问的委任，致委托人

Letter 24

To client, regarding early appointment of consultants

尊敬的先生：

谨建议你方此时委任顾问，以便防止窝工，并保证你方投资获得最大的价值。

根据雇佣条款第2.4条［使用SW/99合同时，替换为“第16条”］，现确认应委任下列顾问进行列出的服务项目：

［列出顾问和服务项目］

如果你方同意，我方将着手与每个公司接洽商谈需要的服务及可支付的费用。委任将通过一系列会议最后确定，到时我方将到会给你建议并说明情况。

我方将十分荣幸接受你方的指示，进行上述事宜。

你忠诚的

Dear Sir

The appointment of consultants is advisable at this stage in order to prevent abortive work and ensure that you obtain the very best value for money.

In accordance with clause 2.4 [*substitute* '*16*' *when using SW/99*] of the terms of engagement, I confirm that the following consultants should be appointed to provide the services noted:

[*List the consultants and services*]

Subject to your agreement, I will approach each firm and negotiate the extent of service required and the fees payable. The appointments can be finalised at a series of meetings at which I will be present to give you general advice and clarification.

I should be pleased to have your instructions to proceed with the course of action outlined above.

Yours faithfully

函件 25

如果委托人询问早期顾问的委任事宜，致委托人

Letter 25

To client, if he queries early appointment of consultants

尊敬的先生：

你方于［填入日期］的来函收悉，深表谢意。从中得知目前你方不想委任顾问。

一座现代化建筑物的拔地而起是一个非常复杂的过程。过去，建筑师可能在不需要其办公室人员以外的任何其他协助的情况下，就能处理一个大型合同项目的各方面问题。但是随着技术、材料和建筑科学的进步，单靠一位专业建筑师已经不可能再包揽所有的工作了。

对这个项目而言，委任顾问是必须的。从长远来看，现阶段委任顾问最为经济，而且你方将从设计工作之初就可以受益于他们的建议。推迟委任顾问不但违背我方的直接建议，并且可能导致项目的窝工。我方将对窝工的工程项目收取金额很高的费用。如果你方能认真考虑此事，并在下周内将你方的指示告知我方，我方将不胜感激。

你忠诚的

Dear Sir

Thank you for your letter of the [*insert date*] and I note that you do not wish to appoint consultants at this stage.

The erection of a modern building is a very complex process. At one time it was possible for an architect to deal with every aspect of a large contract without assistance from anyone other than his own office staff. Advances in technology, materials and building science, however, render such an undertaking impractical for one professional.

Consultants are a necessity on this project. Appointment at this stage will be most economical in the long term and you will have the benefit of their advice from the beginning of the design process. Not only would delay in appointment be against my direct advice, it would result in additional fees chargeable as a result of my abortive work. Such fees would be likely to be substantial. I should be grateful if you would give the matter serious consideration and let me have your instructions within, say, the next week.

Yours faithfully

函件 26

致委托人，随函附上顾问委任草稿

Letter 26

To client, enclosing draft for appointment of consultants

尊敬的先生：

谨提及［填入日期］的系列会议。会上我们与每一方被提议的顾问会面，并最终确认了委任细节。

随函附上下列文件：

1. 我方收到的每一方顾问的协议各两份。我方已审查，并确认与我们会议上商定的条款相符。

2. 我方为你方起草的委任书草稿及寄给每一方顾问的完整的协议。

请确认协议符合你方的要求，并在标明的区域内完成协议，寄给每一方顾问相应的一份副本。剩余的副本留于你处作为参考。

你忠诚的

Dear Sir

I refer to the series of meetings on the [*insert date*] when we met each consultant proposed for this project and we finalised details of the appointment.

I enclose the following:

1. Two copies of the agreement which I have received in respect of each consultant. I have checked them and they correspond with the terms discussed at our meeting.
2. A draft letter which I have prepared for you to send with the completed agreement to each consultant.

Please satisfy yourself that the agreements are in accordance with your requirements, then complete them in the spaces indicated and return one copy of each to the appropriate consultant. The remaining copies should be retained for your own reference.

Yours faithfully

函件 27

委托人致顾问（附有建筑师起草好的委任书草稿）

Letter 27

Client to consultant [*draft prepared by architect*]

尊敬的先生：

我方建筑师已将你方起草的关于此项目的顾问协议一式两份转发给我方。

我方已审查上述文件，并按要求完成了一份副本，现寄于你方。请注意，建筑师将负责与你方协作，将你方的服务综合入整个项目设计。我方希望你方在此方面能全力协作。为避免混淆，除非有十分紧迫的原因，请你方不要直接向我方汇报工作。

你忠诚的

Dear Sir

The architect has forwarded to me two copies of the agreement for consultancy work which you have prepared in respect of the above project.

I have examined the above documents and I return one copy duly completed as requested. Please note that the architect will be responsible for the co-ordination and integration of your services into the overall design and I expect you to co-operate fully in this respect. To avoid confusion, I should be grateful if you would not report directly to me unless there is some very pressing reason.

Yours faithfully

函件 28

如果顾问协议十分复杂，致委托人

Letter 28

To client, if consultants' agreements are complex

尊敬的先生：

谨提及［填入日期］的系列会议。会上我们与各个被提议的顾问方会了面，并讨论了委任细节。

我方已经收到每个顾问方的协议。我方认为这些协议十分复杂，因而需要专家的建议。希望你方能够同意我方以你方的名义与［填入姓名］联系，寻求意见和实际帮助，以确保各顾问方的协议能合理地反映其适当的职责，并且确定各协议之间无任何冲突。

如果你方同意，我方将首先查询相关收费标准，报请你方批准。

你忠诚的

Dear Sir

I refer to the series of meetings on the [*insert date*] when we met each consultant proposed for this project and discussed details of appointment.

I have now received copies of agreements in respect of each consultant. Each agreement is quite complex and, in my view, expert advice is required. I should be pleased to have your agreement to approach [*insert name*] on your behalf to seek advice and practical assistance in ensuring that the agreements to be used for each consultant properly reflect the appropriate duties and that there are no problems with the interaction of one agreement with another.

If you agree, I will obtain details of fee rates for your approval before proceeding.

Yours faithfully

函件 29

如果委托人希望建筑师委任顾问，致委托人

Letter 29

To client, if client wishes architect to appoint consultants

尊敬的先生：

你方于［填入日期］的来函收悉，深表谢意。从中得知你方希望我方为此项目委任顾问。

当然，我方完全可以遵循你方的提议，但这样有违我方的建议。无论以何种方式委任顾问，我方的责任都是负责各方的协作以及综合各顾问方的服务。

顾问费用不受顾问委任方式的影响。但是，在建筑项目中，通常的做法是由委托人自己委任顾问。这样，随着工程的进展，你方能直接接触到其他专业人士。当然，如果出现任何问题也可以直接求助于他们。

我方强烈建议你方再次考虑此事。为避免延误，希望你方能在近几天内给我方回信。

你忠诚的

Dear Sir

Thank you for your letter of the [*inserf dale*] from which I understand that you wish me to appoint the consultants for this project through my office.

It is, of course, perfectly possible to do as you suggest, but it would be against my advice. It is my responsibility to co-ordinate and integrate all consultancy services, however appointed.

Consultancy fees are unaffected by the method of appointment. However, the normal practice in construction projects is that the client appoints his own consultants. By doing so, you have direct access to the other professionals as the work proceeds and, of course, direct recourse to them for any problems which may arise.

I strongly urge you to reconsider and I look forward to hearing from you on this matter within the next few days to avoid delay.

Yours faithfully

函件 30

如果由建筑师来委任顾问，致委托人

Letter 30

To client, if architect is to appoint consultants

尊敬的先生：

你方于［填入日期］的来信收悉，深表谢意。从中得知将由我方代表你方雇佣顾问。此事将即刻着手进行。

你处已有雇佣条款的副本，请你方注意其中第3.11条［使用SW/99合同时，替换为“第17条”］，你方应确认，无论以何种方式委任，每个顾问方应对其履行规定服务的能力和绩效以及对工作的常规检查和执行情况负全部责任。为避免疑问，第3.11条［使用SW/99合同时，替换为“第17条”］将适用于我方按照你方的指示雇佣的所有顾问。

你忠诚的

Dear Sir

Thank you for your letter of the [*insert date*] in which you instruct me that I am to engage the services of consultants on your behalf. The matter will be put in hand.

I should draw your attention to clause 3.11 [*substitute '17' when using SW/99*] of the terms of engagement, a copy of which is already in your possession, which provides that you will hold each consultant however appointed responsible for the competence and performance of the services to be performed by him and for the general inspection and execution of such work. For the avoidance of doubt, clause 3.11 [*substitute '17' when using SW/99*] will apply to all consultants engaged by me on your instructions.

Yours faithfully

函件 31

如果建筑师希望委任顾问，致委托人

Letter 31

To client, if architect wishes to appoint consultants

尊敬的先生：

对于此类项目，我方通常的做法是委任顾问来处理不属于建筑师服务范围的工作。参见雇佣条款的第4.2条［使用SW/99合同时，替换为“第12条”］。

以下为委任的顾问和提供的服务项目以及需支付的额外费用（费用是根据相应的顾问工作收取的），希望你方能够同意。

［列出顾问和服务项目］

请注意第3.11条［使用SW/99合同时，替换为“第17条”］。此条款规定，你方应确认，无论以何种方式委任，每个顾问方对其履行规定服务的能力和绩效以及对工作的常规检查和执行情况负全部责任。为避免疑问，第3.11条［使用SW/99合同时，替换为“第17条”］将适用于我方按照你方的指示雇佣的所有顾问。

你忠诚的

Dear Sir

On projects of this nature, it is my usual practice to appoint consultants to deal with that part of the work not normally within the range of services provided by the architect. Clause 4. 2 [*substitute* '*12*' *when using SW/99*] of the terms of engagement refers.

May I have your agreement to the following appointments and services to be provided and the additional fees payable, such fees to be appropriate to the relevant consultancy.

[*List consultants and services*]

I should draw your attention to clause 3. 11 [*substitute* '*17*' *when using SW/99*] of the conditions which provides that you will hold each consultant however appointed responsible for the competence and performance of the services to be performed by him and for the general inspection and execution of such work. For the avoidance of doubt, clause 3. 11 [*substitute* '*17*' *when using SW/99*] will apply to all consultants engaged by me.

Yours faithfully

函件 32

致顾问，寻求责任免除

Letter 32

To consultant, seeking indemnity

尊敬的先生：

谨提及近期的会议讨论，决定委任你方为上述项目的顾问，负责［写明服务项目］。

在起草正式雇佣合同之前，要求你方出具证据证明你方进行了专业责任保险，并将继续投保此保险。请注明你方在我方委托的工作中，愿意对工作能力、常规检查和绩效等方面出现的种种责任、损失、费用或索赔负全责，并免除我方的任何责任。

你忠诚的

Dear Sir

I refer to the recent discussions regarding your employment as consultant for [*identify the services*] on the above project.

Before a formal contract of engagement can be drawn up, I require you to provide me with proof that you carry and will continue to carry suitable and adequate professional indemnity insurance. Please signify your willingness to indemnify me against any liability, loss, expense or claims of any kind whatsoever in respect of the competence, general inspection and performance of the work entrusted to you.

Yours faithfully

函件 33

如果建筑师提供顾问服务，致委托人

Letter 33

To client if consultancy services to be provided by the architect

尊敬的先生：

对于此类项目，建筑师通常会推荐委任顾问来负责工程的某些方面。我方办公室可提供以下服务，额外的收费符合相关的顾问费用标准。如承蒙你方同意，我方将不胜感激。

［列出提供的顾问服务］

你忠诚的

Dear Sir

On projects of this nature, it would be normal practice for the architect to recommend the appointment of consultants for particular aspects of the work. The following services can be provided by my office and I should be pleased to have your agreement to such provision, and to the additional fees which will be appropriate to the relevant consultancy.

［*List the consultancy services to be provided*］

Yours faithfully

函件 34

致委托人，建议任用具有设计资格的供应商或分包商

Letter 34

To client, suggesting the use of a supplier or sub-contractor in a design capacity

尊敬的先生：

如果你方原则上同意下列专业公司在供应货物和进行项目的部分施工之外，还进行设计工作，我方将十分高兴。我方是按照雇佣条款的第 2.4 条［使用 SW/99 合同时，替换为“第 16 条”］提出此建议的。

［列出专业公司名称和要进行的设计工作］

这种安排在建筑项目中极为常见，因为这些公司在过去的几年里，证实了自己拥有特殊专长或独创的技术体系。在这种情况下，我方认为如果把顾问方的设计工作委托给主承包商进行完成，既不实际，也不划算。当然，交给专业公司的设计工作的费用会从顾问费用中扣除，并且这些公司也会签订一系列相应的担保书及协议，以保障你方利益在设计失误的情况下不受损害。

如果你方同意，我方将起草一份建议总结，提交你方批准。

你忠诚的

Dear Sir

I should be pleased if you would signify your agreement in principle to the use of the following specialist firms to carry out design work in addition to supplying goods and executing work in relation to specific portions of this project. I write in accordance with clause 2.4 [*substitute '16' when using SW/99*] of the terms of engagement.

[*List specialists and the work to be designed by each*]

The arrangement is quite normal in construction projects and generally arises because these firms have a particular expertise or patented systems proven in use over a number of years. In the instances I have in mind it is neither practicable nor economic to commission consultant's designs for execution by the main contractor. The design work in question will, of course, be excluded from consultants' fees and there is a system of warranties and agreements to protect your interests in the event of a design failure.

On receipt of your agreement, I will prepare a summary of my proposals for your approval.

Yours faithfully

函件 35

关于设计、供货和施工公司列表，致委托人

Letter 35

To client, regarding lists of firms to design and supply/execute

尊敬的先生：

谨提及我方于［填入日期］的信函及你方于［填入日期］的关于批准任用具有设计资格的专业公司的回函。

我方建议邀请以下公司对标明的工程项目部分进行投标：

［列出项目部分及建议的公司］

请你方给予批准或批示，以便我方立即着手准备必要的文件。

你忠诚的

Dear Sir

I refer to my letter of the [*insert date*] and your reply of the [*insert date*] approving the use of specialist firms in a design capacity.

I propose that the following firms be invited to tender for the parts of the project as indicated:

[*List parts of project and the firms proposed*]

Please let me have your approval or observations so that I can begin the preparation of the necessary documentation without delay.

Yours faithfully

函件 36

关于供应商公司列表，致委托人

Letter 36

To client, regarding lists of firms to supply

尊敬的先生：

谨建议邀请下列公司仅对于以下货物供应进行投标，如果你方能同意，我方将不胜感激。

［列出货物种类及建议的投标公司］

请你方给予批准或批示，以便我方立即着手准备必要的文件。

你忠诚的

Dear Sir

I should be pleased if you would signify your agreement to the following list of firms which I propose to invite to tender for the supply only of goods as follows:

[*List goods and the firms proposed*]

Please let me have your approval or observations so that I can begin to prepare the necessary documentation without delay.

Yours faithfully

函件 37

如果建议使用较新的材料或工艺，致委托人

Letter 37

To client, if a relatively new material or process is proposed

尊敬的先生：

谨通过书面形式确认我们近期关于［填入材料或工艺的名称］的谈话。

尽管此材料/工艺［视情况取舍］仅仅投入使用了［填入一段时间］，但到目前为止非常成功。我方已经以你方名义致函制造商，说明了我们计划使用此材料/工艺［视情况取舍］，他们也给予了积极的答复。他们已经致函我方表示愿意签署一份对你方有利的担保书。我方还收到了［填入适当的国内或国际组织的名称］对该公司的好评和随此函附上的他们所有的往来函件的副本。

使用未经试验的新材料或工艺免不了会有风险，因而将由你方做最后决定。不过我方相信这种新材料/工艺［视情况取舍］的风险是在可接受的限度之内。而且没有其他的材料/工艺［视情况取舍］有这么多的优点。请你方在［填入日期］之前做出决定。如果你方就此事需要进一步的专业建议，请立即电话告知我方。

你忠诚的

Dear Sir

I am writing to confirm our recent conversation regarding the use of [*insert name of material or process*].

The material/process [*delete as appropriate*] has only been in use for [*insert period*], but so far it appears to be successful. I have written to the manufacturers on your behalf explaining the proposed use and, as might be expected, they are entirely reassuring. They have put that reassurance in the form of a letter and expressed their willingness to complete a form of warranty in your favour. I have also received favourable comments from [*insert names of appropriate national or international technical organisations*] and copies of all correspondence are enclosed with this letter.

There is always a risk in using relatively untried materials or processes and, therefore, the final decision must be yours. However, I believe that the risk is within acceptable limits in this instance and there is no other material/process [*delete as appropriate*] which offers so many advantages. I should be pleased to have your decision not later than [*insert date*]. If you require any further advice on the matter, please do not hesitate to telephone me.

Yours faithfully

函件 38

致设计团队成员，安排会议

Letter 38

To members of the design team, arranging meeting

尊敬的先生：

谨通知于［填入日期和时间］在我处办公室/工地［视情况取舍］召开设计团队会议。请届时参加会议，我方将深感荣幸。

［视情况加上以下一条或多条：］

会议预期要进行［填入一段时间］。
会议提供汉堡。
随函附上会议议程。

你忠诚的

Dear Sir

There will be a meeting of the design team for the above project at [*insert time*] on [*insert date*] at this office/on site [*delete as appropriate*]. I should be pleased if you would arrange to attend.

[*Add one or more of the following sentences as appropriate*:]

The meeting will be expected to last [*insert period*].

Sandwiches will be provided.

An agenda is attached to this letter.

Yours faithfully

函件 39

关于规划审批的费用，致委托人

Letter 39

To client, regarding fees for planning applications

尊敬的先生：

我方预计于［填入日期］进行初步/完全［视情况取舍］规划许可的申请工作。所有的规划管理部门都要按照收费明细表收取费用。

我方已经按有关部门公布的收费标准计算了规划审批需要支付的费用，总计［填入金额］。希望你方能按此金额开具以［填入规划管理部门的名称］为受款人的支票。规划机构在收到申请书并核查了提交的文件后，最终确认交费金额。因此如果规划机构的收费超出我方的预算，你方需补交不足费用。规划机构只有在收到足额的费用之后才会处理申请事宜。

你忠诚的

Dear Sir

I anticipate being in a position to make application for outline/full [*delete as appropriate*] planning permission on the [*insert date*]. All planning authorities are required to charge fees in accordance with a schedule of charges.

I have calculated the fee payable in respect of this application from the authority's published scale of charges as [*insert amount*] and I should be pleased to receive your cheque for this amount which you should draw in favour of [*insert name of planning authority*]. Final confirmation of the fees due will only be made by the authority when they have received the application and checked the submitted documents. Therefore, if the authority disagrees with my calculations you may be required to pay an additional amount. The application will not be considered until the correct fee is received by the authority.

Please let me know if you require any further information.

Yours faithfully

函件 40

致规划管理部门，申请批准初步规划

Letter 40

To planning authority, requesting outline approval

尊敬的先生：

谨提及我方与你方［填入姓名］对此提案的讨论，现代表我方的委托人［填入姓名］向你方提交批准初步规划的申请书。

随函附上：

1. 签好的申请表 3 份。
2. 标明场地和选址的编号为［填入编号］的图纸 3 份。
3. 编号为［填入你想提交的其他图纸编号］的图纸各 3 份。
4. 费用金额［视情况填入金额。如果无须交纳费用，删去此项］。

［视情况，加上：］

我方现确认随函附上的公告曾于［填入日期］在［视情况填入地区报纸名称］上发布。此广告曾在工地明显位置展示了 7 天［填入起止日期］。

如果申请书的内容出现任何问题，请你方打电话通知我方，以便尽快解决。
对此我方将不胜感激。

你忠诚的

Dear Sir

I refer to my discussions with your [*insert name*] with regard to this proposal and I now formally submit an application for outline planning approval on behalf of my client [*insert name*].

I enclose:

1. Three copies of the form of application duly signed.
2. Three copies of drawing number [*insert number*] showing the site and its location.
3. Three copies of drawings numbers [*insert numbers of any other drawings which you wish to include*].
4. The fee of [*insert amount of fee as appropriate. If no fee is payable, delete this item*].

[*Add, if appropriate*:]

I hereby certify that the enclosed advertisement has been published on the [*insert dates*] in the [*insert name of local newspaper and add, if appropriate*:] and the said advertisement has been displayed in a prominent position on the site for a period of 7 days from [*insert date*].

If there are any points arising out of this application, I should appreciate a telephone call to resolve them as quickly as possible.

Yours faithfully

函件 41

如果委托人希望不等获得必要的批准就继续进行工作，致委托人

Letter 41

To client, if client considering if work should proceed before necessary approvals obtained

尊敬的先生：

我们在近期关于工程进度的讨论中，谈到如果不等官方正式的［填入类别］批准就进行设计工作，就可以节省［填入一段时间］。

我方仔细研究了此事，正式的批准应于［填入日期］下达。如果到下达批准之日的所有设计工作窝工，从［填入日期］到［填入日期］的总共专业费用大约为［填入金额］。

当然，这样做的风险在于，申请书也许不能获得批准，部分甚至全部的设计工作也许变得徒劳。由于我方可能对有些方面的因素不了解，此事需要你方根据具体情况斟酌决定。如果你方想继续进行设计工作，就请尽快做出决定。如能在［填入日期］之前收到你方的书面指示，我方将不胜感激。

你忠诚的

Dear Sir

I refer to our recent discussion regarding the progress of this project, when it was suggested that [*insert period*] could be saved by proceeding with the design development without waiting for formal [*insert type*] approval.

I have looked into the matter and the position is that formal approval should be given on [*insert date*]. If all the design work carried out up to that date proves to be abortive, the total professional fees incurred between [*insert date*] and [*insert date*] would be approximately [*insert amount*].

There is, of course, a risk that approval may be refused and some if not all the design development work may be wasted. It is a matter for you to judge in the light of your own particular circumstances which may involve considerations of which I am unaware. If design work is to proceed, there is some urgency to your decision and I will be grateful to receive your written instructions by [*insert date*].

Yours faithfully

函件 42

致规划管理部门，要求更新临时许可证

Letter 42

To planning authority, requesting renewal of temporary permission

尊敬的先生：

谨提及你方于［填入日期］签发的关于上述项目的临时规划许可证［填入号码］。现代表我方的委托人［填入姓名］正式申请延长临时许可证的期限至［填入日期］。

［视情况加上：］

此外［说明认为能够促使规划管理部门延长许可期限的事项］。

如果申请书的内容出现任何问题，请你方打电话通知我方，以便尽快解决。我方将不胜感激。

你忠诚的

Dear Sir

I refer to temporary planning permission number [*insert number*] granted on the [*insert date*] in respect of the above development. I now formally apply on behalf of my client [*insert name*] for an extension of the duration of the temporary permission until [*insert date*].

[*Add, if appropriate:*]

In support of this application [*state the matters which you consider may influence the authority to extend the duration of the permission*].

If there are any points arising out of this application, I should appreciate a telephone call in order to resolve them as speedily as possible.

Yours faithfully

Copy: Client

函件 43
致委托人，随函附上费用账目

Letter 43
To client, enclosing fee account

尊敬的先生：

按照雇佣条款表 3 ［使用 SW/99 合同时，删去“表 3”］，现随函附上我方的［填入编号］费用账目，其上注明了我方截止到［填入日期］的费用支出情况。

你方如能马上付款，我方将不胜感激。

你忠诚的

Dear Sir

In accordance with the terms of engagement, Schedule Three [*omit 'Schedule Three' when using SW/99*], I enclose my fee account [*insert number*] which includes a note of my expenses to [*insert date*].

Prompt payment would be appreciated.

Yours faithfully

函件 44

如果委托人推迟支付费用，致委托人：第一次提醒

Letter 44

To client, if fees are late: first reminder

尊敬的先生：

我方注意到从［填入日期］发出［填入编号］费用账目后，仍然没有收到你方付款。

无疑，你方并未注意到此事。希望你方能在几天内将支票寄给我方。在当今金融紧缩的情况下，要使费用保持在合理水平的惟一方法就是即时付款。

你忠诚的

Dear Sir

I note that my fee account number [*insert number*] of the [*insert date*] is still outstanding.

No doubt the matter has escaped your attention, but I should be pleased if you would let me have your cheque within the next few days. The only way that fees can be kept at a reasonable level in these times of financial stringency is by securing prompt payment.

Yours faithfully

函件 45

如果委托人推迟支付费用，致委托人：第二次提醒

Letter 45

To client, if fees are late: second reminder

尊敬的先生：

谨提及我方于［填入日期］发出的［填入编号］费用账目以及我方于［填入日期］要求你方即刻付款的信函。

很遗憾，到我方写信之时，仍然没有收到你方的支票。希望你方立即处理此事，以避免不必要的行政手续费用。

你忠诚的

Dear Sir

I refer to my fee account number [*insert number*] of the [*insert date*] and my letter of the [*insert date*] in which I requested prompt payment.

I regret that at the time of writing I have not received your cheque and I should be pleased if you would treat this matter with some urgency to avoid unnecessary administrative costs.

Yours faithfully

函件 46

如果委托人推迟支付费用，致委托人：第三次提醒

Letter 46

To client, if fees are late: third reminder

尊敬的先生：

尽管我方于［填入日期］两次给你方寄发了提醒信函，我方于［填入日期］发出的［填入编号］费用账目仍然未收到付款。

鉴于我方一直认为我们之间的工作关系良好，因而没有严格追究此事。尽管我方不想给你方造成任何麻烦，但是我方必须考虑自身的经济利益。

可否请你方立即回寄足额的支票？

你忠诚的

Dear Sir

I refer to my fee account number [*insert number*] of the [*insert date*] which has not yet been paid despite reminders sent to you on the [*insert dates*].

In view of what I always took to be the good working relationship which exists between us, I have not pursued this matter with the vigour it deserves. Although I have no desire to cause problems for you, I must have a care for my own financial position.

May I expect a cheque for the full amount by return of post?

Yours faithfully

函件 47

如果委托人推迟支付费用，致委托人：警告采取法律诉讼

Letter 47

To client, if fees are late: legal action threatened

尊敬的先生：

我方已经于［填入几个日期］几次向你方寄发了提醒信函，但我方发出的［填入编号］费用账目仍然没有收到你方的付款。因此，如果到［填入日期，此信日期7天后］仍然未收到你方第一时间寄来的支票，我方将会很遗憾地告诉你方，我方不得不立即采取法律手段要求偿付。我方将选择裁决、仲裁或者视情况选择其他法律程序。

［视情况，加上：］

届时，我方会利用我方的职权停止该项目的所有工作，直到收到付款为止。

你忠诚的

Dear Sir

I refer to my fee account number [*insert number*] of the [*insert date*] in the sum of [*insert amount*].

I have sent you reminders on the [*insert dates*], but I have not yet received payment. I regret, therefore, that if I do not receive your cheque for the full amount by first post on the [*insert date, which should be 7 days after the date of this letter*], I shall have no alternative but to take immediate action for recovery. This may be by way of adjudication, arbitration or legal proceedings as I deem to be appropriate.

[*Add, if desired and appropriate*:]

At that time, I will exercise my right to stop all work on this project until I receive full payment.

Yours faithfully

第三章　初步建筑方案

把建筑师的工作阶段分割得像防水舱一样完全是人为的界定。实际上，建筑师的各项工作经常会前后关联，很多事情进行得要比《工作计划》中设想的要早或晚。但无论如何，分成各阶段是对工程进度的一个有益的制约。在C阶段，建筑师将忙于工程基本策划，几乎不需要用到信函，无论是标准信函或是其他信函。

本阶段可能出现的问题有：你的初步规划的批准申请遭到拒绝。市容协会对所有的规划申请都严格审查，甚至有可能在你还没构思出该项目的具体形状之前，就已经面临着对自己辛勤努力的强烈反对。但有时你的一封礼貌而又合作的回信就可以应付这种难题，而市容协会也可能会转而支持你。

这时大概是你能够让你的委托人正式确定合同条款的最早时机。委托人往往无法在这时做最终决定，而要等到F阶段末期。不要忘记你应该有足够的法律知识来建议你的委托人使用哪一种合同格式。如果因为你的错误建议而造成委托人一方的损失，你可能要承担责任。

如果合同格式为JCT设计施工合同（WCD98），在此阶段你要与委托人就《雇主要求》的内容达成一致意见。

函件 48

如果市容协会来函表示反对，致委托人

Letter 48

To client, if amenity society write with objections

尊敬的先生：

谨提及我们［填入日期］的电话谈话，涉及到市容协会对规划许可申请的反对。现随函附上他们于［填入日期］的来函以及我方于［填入日期］的回函副本。

我方的信函涵盖了我们讨论过的所有事项。希望我方与他们见面时，能够解除他们的顾虑。

你忠诚的

Dear Sir

I refer to our telephone conversation of the [*insert date*] when we discussed the objections of the amenity society and I enclose a copy of their letter dated [*insert date*] and my reply dated [*insert date*].

My letter covers the points we discussed and I hope I can allay their fears when I see them.

Yours faithfully

函件 49

如果他们来函表示反对，致市容协会

Letter 49

To amenity society, if they write with objections

尊敬的先生：

你方于［填入日期］的来函收悉，深表谢意。我方理解你方来函的原因。我方认为该工程项目就其与周围建筑的位置关系来说完全符合所有的合理要求。不仅如此，此项目建成后，还能够很大程度地促进周围环境。

不过，我方决定听取尽可能多的正反两方面的意见。最好的方法就是参加你方的成员会议，听取他们的意见，并详细解释此规划书，解答他们的疑问。

我方将再次致函商定合适的时间。

你忠诚的

抄送：委托人

Dear Sir

Thank you for your letter of the [*insert date*]. I understand and appreciate your reasons for writing. Naturally, I consider this project to satisfy all reasonable requirements in respect of its position in relation to neighbouring buildings and I am confident that it will do more and make a serious contribution to the environment.

However, I am determined to gather as many views as possible, favourable or otherwise, and I suggest that this can best be accomplished if I attend a meeting of your members to listen to them, and to explain the scheme in detail and answer questions.

I will write to you again to arrange a convenient date.

Yours faithfully

Copy: Client

函件 50a

关于使用的合同格式，致委托人

此函件仅适用于 JCT98 合同

Letter 50a

To client, regarding form of contract to be used

This letter is only suitable for use with JCT 98

尊敬的先生：

根据我们［填入日期］的会议，我方认为有必要确定下列达成的初步决定：

［视情况取舍下列各项可选方案］

1. 使用的合同格式为 1998 年 JCT 施工合同标准格式/地方政府/私人/计算工程量/不计算工程量/大致计算工程量/及其合同修正案前四条。

2. 此格式将按下列事项完成：

- 第二项陈述中，删去对分项工程预算表的参考。
- 第六项陈述将被删去。
- 本合同将作为契约合同/非盖印合同执行。
- 第 5.3 条：要求/不要求承包商提交总进度计划/在网络分析或优先图表中需要总体计划。
- 建议为 21.2.1 条的工程保险做好准备。
- 第 22A/22B/22C 条适用本合同。
- 要求第 22D 条的工程保险。
- 随函附上按商定事项完成的附件一份。

3. 使用“分段完工补充条款”。

4. 使用“承包商设计部分补充条款”。

请仔细核查此信函内容。如果发现我方对你方的指示理解有误或者你方对任何事项有疑问，请立即告知我方。

你忠诚的

抄送：工料测量师

（一般而言，无论在英国或在北美，工程咨询方包括建筑师（Architect）/工程师（Engineer），即 A/E，以及造价工程师（CE：Cost Engineer）或工料测量师（QS：Quantity Surveyor），其中北美实行造价工程师制，强调工程师背景，而英国或英联邦国家则称工料测量师，强调其独立的执业资格——译者注。）

Dear Sir

Following our meeting on the [*insert date*] I thought it would be useful to confirm some preliminary decisions reached as follows:

[*Delete as appropriate from the following alternatives*:]

1. The contract to be used is the JCT Standard Form of Building Contract/Local Authorities/Private/With Quantities/Without Quantities/With Approximate Quantities/1998 with Amendments up to and including Amendment 4.
2. The form will be completed as follows:
- In the second recital, reference to the priced activity schedule will be deleted.
- The sixth recital will be deleted.
- The contract will be executed as a deed/under hand.
- Clause 5.3: a master programme is/is not required from the contractor/it will be required in network analysis or precedence diagram form.
- It will be advisable to make provision for clause 21.2.1 insurance.
- Clause 22A/22B/22C is applicable.
- Clause 22D insurance may be required.
- A copy of the appendix is enclosed, completed as agreed.
3. The Sectional Completion Supplement will be used.
4. The Contractor's Designed Portion Supplement will be used.

Please check the contents of this letter carefully and let me know immediately if I have misunderstood your instructions or if there is any point on which you require further clarification.

Yours faithfully

Copy: Quantity surveyor

函件 50b

关于使用的合同格式，致委托人

此函件仅适用于 IFC98 合同

Letter 50b

To client, regarding form of contract to be used

This letter is only suitable for use with IFC 98

尊敬的先生：

根据我们［填入日期］的会议，我方认为有必要确定下列达成的初步决定：

［视情况取舍下列各项可选方案］

1. 使用的合同格式为 1998 年 JCT 中等规模施工合同标准格式，包括合同修正案前四条。

2. 此格式将按下列事项完成：

- 第一项陈述：删去“建筑师”/“合同管理人”。删去“工程详细说明”/“工作进度”/“工程量清单”。删去最后一段。
- 第二项陈述：删去可选方案 A/B。
- 删去第四项陈述。
- 删去/保留斜体文字。
- 第四款：删去/保留斜体文字。
- 本合同将作为契约合同/非盖印合同执行。
- 第七款：删去可选方案 A/B。
- 建议为第 6. 2. 4 条的工程保险做好准备。
- 第 6. 3A/6. 3B/6. 3C 条适用本合同。
- 要求第 6. 3D 条的工程保险。
- 随函附上按商定事项完成的附件一份。

请仔细核查此信函内容。如果发现我方对你方的指示理解有误或者你方对任何事项有疑问，请立即告知我方。

你忠诚的

抄送：工料测量师

Dear Sir

Following our meeting on the [*insert date*] I thought it would be useful to confirm some preliminary decisions reached as follows:

[*Delete as appropriate from the following alternatives*:]

1. The contract to be used is the JCT Intermediate Form of Building Contract 1998 with Amendments up to and including Amendment 4.
2. The form will be completed as follows:

- First recital: 'the Architect'/'the Contract Administrator' will be deleted. Delete 'the Specification'/'the Schedules of Work'/'Bills of Quantities'. Delete the final paragraph.
- Second recital: delete alternative A/B.
- Delete the fourth recital.
- The words in italics will be struck out/left in.
- Article 4: the words in italics will be struck out/left in.
- The contract will be executed as a deed/under hand.
- Article 7: alternative A/B will be deleted.
- It will be advisable to make provision for clause 6. 2. 4 insurance.
- Clause 6. 3A/6. 3B/6. 3C is applicable.
- Clause 6. 3D insurance may be required.
- A copy of the appendix is enclosed, completed as agreed.

Please check the contents of this letter carefully and let me know immediately if I have misunderstood your instructions or if there is any point on which you require further clarification.

Yours faithfully

Copy: Quantity surveyor

函件 50c

关于使用的合同格式，致委托人

此函件仅适用于 MW98 合同

Letter 50c

To client, regarding form of contract to be used

This letter is only suitable for use with MW 98

尊敬的先生：

根据我们［填入日期］的会议，我方认为有必要确定下列达成的初步决定：

［视情况取舍下列各项可选方案］

1. 使用的合同格式为 1998 年 JCT 小型建筑施工协议，包括协议修正案前四条。
2. 此格式将按下列事项完成：

- 第一项陈述：删去“建筑师”／“合同管理人”。删去“图纸……合同图纸”／“工程详细说明……合同详细说明”／“明细表”。
- 第二项陈述：删去“工程详细说明”／“或明细表”／“或提供费率明细表”。
- 第四项陈述：删去。
- 第五项陈述：删去可选方案 A/B。
- 第三款：删去“建筑师”／“合同管理人”。
- 第四款：删去可选方案 A/B/全部删去此款。
- 第五款：全部删去此款。
- 第六款：删去英国皇家建筑师协会/皇家特许测量师学会/建筑联盟/国家专业承包商理事会。
- 第 7A 款：删去英国皇家建筑师协会/皇家特许测量师协会/皇家特许仲裁员协会/全部条款。
- 第 7B 款：全部删去此款。
- 本合同将作为契约合同/非盖印合同执行。
- 第 2.1 条：开工日期为［填入日期］，完工日期为［填入日期］。
- 第 2.3 条：预定违约赔偿金为每［填入“天”或“周”］［填入金额］。
- 第 2.5 条：服务缺陷责任期为［填入一段时间］。
- 第 3.6 条和第 4.1 条：根据以下陈述，视情况删去。
- 第 4.5 条：期限为［填入一段时间］。
- 第 4.6 条：删/不删此条。

（转下页）

Dear Sir

Following our meeting on the [*insert date*] I thought it would be useful to confirm some preliminary decisions reached as follows:

[*Delete as appropriate from the following alternatives*:]

1. The contract to be used is the JCT Agreement for Minor Building Works 1998 with Amendments up to and including Amendment 4.
2. The form will be completed as follows:

- First recital: 'the Architect' / 'the Contract Administrator' will be deleted. Delete 'drawings … Contract Drawings?)' / 'a Specification … Contract Specification?)' / 'schedules'.
- Second recital: delete 'the Specification' / 'or the schedules' / 'or provided a schedule of rates'.
- Fourth recital: delete.
- Fifth recital: delete alternative A/B.
- Article 3: 'the Architect' / 'the Contract Administrator' will be deleted.
- Article 4: delete alternative A/B/whole of this article.
- Article 5: delete the whole of this article.
- Article 6: delete Royal Institute of British Architects/Royal Institution of Chartered Surveyors/Construction Federation/National Specialist Contractor's Council.
- Article 7A: delete Royal Institute of British Architects/Royal Institution of Chartered Surveyors/Chartered Institute of Arbitrators/the whole of this article.
- Article 7B: delete the whole of this article.
- The contract will be executed as a deed/under hand.
- Clause 2.1: the date for commencement will be [*insert date*] and the date for completion will be [*insert date*].
- Clause 2.3: the liquidated damages will be [*insert amount*] per [*insert 'day' or 'week'*].
- Clause 2.5: the defects liability period will be [*insert period*].
- Clauses 3.6 and 4.1: deletions as appropriate to follow recitals.
- Clause 4.5: the period will be [*insert period*].
- Clause 4.6 will/will not be deleted.

[*continued*]

函件 50c 续表

Letter 50c continued

- 第 5. 2 条：承包商公布后完成此条。
- 第 6. 2 条：金额为［填入金额］。
- 第 6. 3A 条：此条适用本合同，专业费用的百分比为［填入百分比］。
- 第 6. 3B 条适用本合同。

请仔细核查此信函内容。如果发现我方对你方的指示理解有误或者你方对任何事项有疑问，请立即告知我方。

你忠诚的

抄送：工料测量师［如果已经委任］

- Clause 5. 2 will be completed when the contractor is known.
- Clause 6. 2: the amount shall be [*insert amount*].
- Clause 6. 3A will apply and the percentage to cover professional fees will be [*insert percentage*].
- Clause 6. 3B will apply.

Please check the contents of this letter carefully and let me know immediately if I have misunderstood your instructions or if there is any point on which you require further clarification.

Yours faithfully

Copy: Quantity surveyor [*if appointed*]

函件 50d

关于使用的合同格式，致委托人

此函件仅适用于 WCD98 合同

Letter 50d

To client, regarding form of contract to be used

This letter is only suitable for use with WCD 98

尊敬的先生：

根据我们［填入日期］的会议，我方认为有必要确定下列达成的初步决定：

［视情况取舍下列各项可选方案］

1. 使用的合同格式为 1998 年 JCT 承包商设计施工合同标准格式，包括合同修正案前四条。

2. 此格式将按下列事项完成：

- 本合同将作为契约合同/非盖印合同执行。
- 建议为第 21.2.1 条的工程保险做好准备。
- 第 22A/22B/22C 条适用本合同。
- 要求第 22D 条的工程保险。
- 删去第 30.4.2.2 条，拟定一条代替条款，表明各方不想将保留金存于单独的银行账户上。
- 补充条款 S2/S3/S4/S5/S6/S7 适用本合同。

3. “分段完工修订案”适用于本合同。

4. 随函附上按商定事项完成的附件 1、2、3 各一份。

请仔细核查此信函内容。如果发现我方对你方的指示理解有误或者你方对任何事项有疑问，请立即告知我方。

你忠诚的

抄送：工料测量师［如果已经委任］

Dear Sir

Following our meeting on the [*insert date*] I thought it would be useful to confirm some preliminary decisions reached as follows:

[*Delete as appropriate from the following alternatives*:]

1. The contract to be used is the JCT Standard Form of Building Contract With Contractor's Design 1998 with Amendments up to and including Amendment 4.
2. The form will be completed as follows:

- The contract will be executed as a deed/under hand.
- It will be advisable to make provision for clause 21.2.1 insurance.
- Clause 22A/22B/22C insurance is applicable.
- Clause 22D insurance may be required.
- Clause 30.4.2.2 will be deleted and a replacement clause will be drafted to express the intentions of the parties not to place the retention in a separate banking account.
- Supplementary provisions S2/S3/S4/S5/S6/S7 will apply.

3. The sectional completion modifications will apply.
4. A copy of appendices 1, 2 and 3 are enclosed, completed as agreed.

Please check the contents of this letter carefully and let me know immediately if I have misunderstood your instructions or if there is any point on which you require further clarification.

Yours faithfully

Copy: Quantity surveyor [*if appointed*]

函件 50e

关于使用的合同格式，致委托人

此函件仅适用于GC/Works/1（1998）合同

Letter 50e

To client, regarding form of contract to be used

This letter is only suitable for use with GC/Works/1 (1998)

尊敬的先生：

根据我们［填入日期］的会议，我方认为有必要确定下列达成的初步决定：

［视情况取舍下列各项可选方案］

1. 使用的合同格式为GC/Works/1（1998）标准政府合同格式，计算工程量/不计算工程量。

2. 此格式将按下列事项完成：

- 第8.2条中，为雇主责任投保的最小金额为［填入金额］。
- 第8.3条中，要求有可选方案A/B/C。
- 第8.3（a）条中，专业费用的百分比为［填入百分比］。
- 第8A条适用/不适用本合同。
- 要求承包商承担工程以下部分的设计工作：［填入简要描述］。要求有可选方案A/B。
- 合同期限为［填入一段时间］。
- 本合同将作为契约合同/盖印合同/非盖印合同执行。
- 随函附上一份按商定事项完成的条款详项的摘要。

请仔细核查此信函内容。如果发现我方对你方的指示理解有误或者你方对任何事项有疑问，请立即告知我方。

你忠诚的

抄送：工料测量师［如果已经委任］

Dear Sir

Following our meeting on the [*insert date*] I thought it would be useful to confirm some preliminary decisions reached as follows:

[*Delete as appropriate from the following alternatives*:]

1. The contract to be used is the Standard Government Form of Contract GC/Works/1 (1998) With Quantities/Without Quantities.
2. The following should be noted:
- The minimum amount insured in respect of employer's liability in clause 8(2) is £ [*insert amount*].
- In respect of clause 8(3), alternative A/B/C is required.
- The percentage of professional fees in clause 8(3)(a) is [*insert percentage*].
- Clause 8A will/will not apply.
- The contractor is required to undertake the design of the following part of the Works: [*insert brief description*]. Alternative A/B is required.
- The contract period will be [*insert period*].
- The contract will be executed as a deed/by sealing/under hand.
- A copy of the abstract of particulars is enclosed completed as agreed.

Please check the contents of this letter carefully and let me know immediately if I have misunderstood your instructions or if there is any point on which you require further clarification.

Yours faithfully

Copy: Quantity surveyor

函件 51

关于《雇主要求》的内容，致委托人

此函仅适用于 WCD98 合同

Letter 51

To client, regarding the content of the Employer's Requirements

This letter is only suitable for use with WCD 98

尊敬的先生：

准备《雇主要求》是我方的职责，它将组成合同的一部分。《雇主要求》是非常重要的文件，因为《承包商建议书》必须符合其规定。因此《雇主要求》必须包括所有相关事宜。《雇主要求》的内容必须花一定的时间来商定，我方认为最好制定出议程以保证最充分地利用时间。希望你方能在会面之前认真考虑以下事项。我方将在你细阅此函后几天内，给你打电话。

[列出要考虑的事项，可能包括以下部分或所有的条目

- 场地及边界问题细节。
- 服务设施要求细节。
- 建筑物的用途。
- 其他任何可能影响到准备《承包商建议书》及其价格的事项。
- 功能需求或辅助性需求说明：
  ——建筑物的数量和种类
  ——房屋密度及组合
  ——高度限制
  ——竣工的具体要求。
- 预算金额。
- 规划限制。
- 规定或其他契约。
- 法定的或其他许可。
- 场地要求。
- 《雇主要求》的法定范畴。
- 出入限定条件。
- 公共设施情况。
- 《承包商建议书》表述。
- 承包商图纸提交。

（转下页）

Dear Sir

It is part of my duty to prepare the Employer's Requirements which will form part of the contract. It is an important document, because it is what the Contractor's Proposals must satisfy. Therefore, it must include all relevant matters. It is essential that we spend some time discussing the content and I believe that the best way to ensure maximum use of time is to have an agenda. Perhaps you would give the following items some thought before we meet. I will telephone you within the next few days after you have had the opportunity to digest the contents of this letter:

[*List the matters which require consideration. They may include some or all of the following*:

- *Details of site and boundaries.*
- *Details of accommodation requirements.*
- *Purposes for which the building is to be used.*
- *Any other matter likely to affect the preparation of the Contractor's Proposals and his price.*
- *Statement of functional and ancillary requirements*:
  - *kind and number of buildings*
  - *density and mix of dwellings*
  - *height limitations*
  - *specific requirements for finishes.*
- *Provisional sums.*
- *Planning constraints.*
- *Restrictive or other covenants.*
- *Statutory or other permissions.*
- *Site requirements.*
- *The mandatory extent of the Employer's Requirements.*
- *Access restrictions.*
- *Availability of public utilities.*
- *Presentation of Contractor's Proposals.*
- *Submission of contractor's drawings.*

[*continued*]

函件 51 续表

Letter 51 continued

- 详细建造图纸的要求。
- 分阶段或分期付款。
- 雇主代理人的职责。
- 完成合同附件的必需信息。]

你忠诚的

抄送：工料测量师［如果已经委任］

- *Detailed as-built drawings requirements.*
- *Stage or periodic payments.*
- *Functions of/the employer's agent.*
- *Information required to complete the contract appendices.*]

Yours faithfully

Copy: Quantity surveyor [*if appointed*]

函件 52
致委托人，随函附上项目初步建筑方案报告书

Letter 52
To client, enclosing outline proposals report

尊敬的先生：

我方已经完成对此项目的初步建筑方案。现欣然附上报告［填入份数］份，以及图解此项目方案的编号为［填入编号］的图纸数份。

现确认我方将于［填入日期］［填入时间］去你处，讨论我方的报告，并听取你方的指示。

［视情况，加上：］

请你方特别注意［视情况，填入事项］。

你忠诚的

Dear Sir

I have completed my outline proposals for this project and I have pleasure in enclosing [*insert number*] copies of my report and drawings numbers [*insert numbers*] illustrating the scheme.

I confirm that I will visit you on the [*insert date*] at [*insert time*] to discuss my report and to receive your instructions.

[*If appropriate, add:*]

Will you give special consideration to [*insert as appropriate*].

Yours faithfully

第四章　详细建筑方案

在此阶段，建筑师将最终完成自己的设计，并解决与相关政府部门之间的问题。这一阶段的末期，明智的建筑师会做好一份简要的报告，连同展示图纸一起提交给委托人。但你必须说明，此后对雇主要求的任何变更将延误项目进程，并且你会对此另行收费。

函件 53

致制造商，索取技术说明材料

Letter 53

To manufacturer, asking for technical literature

尊敬的先生：

欣然希望你方能寄来［填入产品名称］的完整的技术细节说明。

请注意我方不希望此时与你方代表见面。如果你方的说明材料表明你方产品符合我方的要求，我方会及时要求你方的进一步协助。

你忠诚的

Dear Sir

I should be pleased to receive full technical details of [*insert name of product*].

Please note that I do not wish to meet your representative at this stage. If your literature indicates that your product is appropriate to my requirements, I may request further assistance in due course.

Yours faithfully

函件 54

致制造商，要求派代表进行一般性来访

Letter 54

To manufacturer, asking representative to visit for general purposes

尊敬的先生：

我方对你方的［填入产品名称］很感兴趣。如果你方能在［填入一段时间］内派技术代表电话联系我方，约定见面事宜，我方将深感荣幸。

请注意，要求该技术代表能够提供成功使用你方产品的工程项目细节及情况说明书，以便我方备案。

你忠诚的

Dear Sir

I am interested in [*insert name of product*] and I should be pleased if you would ask your technical representative to telephone to make an appointment to see me within [*insert time period*].

Please note that he must be prepared to supply details of projects on which your product has been used successfully and fact sheets for my office library.

Yours faithfully

函件 55

致制造商，要求派代表进行专门项目的来访

Letter 55

To manufacturer, asking representative to visit for special purposes

尊敬的先生：

我方打算在新项目上使用［填入产品名称］，现需要技术支持。

请安排你方技术代表电话联系我方，约定在我方办公室/工地［视情况取舍］面谈事宜。要求该技术代表能够提供准确、翔实的技术资料及成功使用你方产品［填入产品名称］的项目列表。

你忠诚的

Dear Sir

I am considering the use of [*insert name of product*] on a new project and I require some technical advice.

Please arrange for your technical representative to telephone to make an appointment to visit me at this office/meet on site [*delete as appropriate*]. He should be prepared to supply precise and substantiated facts and a list of projects where [*insert name of product*] has been used successfully.

Yours faithfully

函件 56

致制造商，索要确认信函

Letter 56

To manufacturer, asking for letter

尊敬的先生：

谨提及我们于［填入日期］的会议上，讨论了此项目中［填入产品名称］的使用事宜。如果你方能致函我方，明确确认［填入产品名称］将符合我方委托人的要求，我方将明确规定使用该产品。谨确认我们已经详细讨论了这些要求，所有要求现总结如下：

［简洁明了地总结委托人关于此产品的要求］

我方期待最迟于［填入日期］之前收到你方来函。

你忠诚的

Dear Sir

Following our meeting on the [*insert date*] to discuss the use of [*insert name of product*] on this project, I am prepared to specify its use if you will write to me unequivocally confirming that [*insert name of product*] will satisfy my client's requirements. I confirm that we discussed those requirements in detail, but they may be summarised non-exclusively as [*summarise the client's requirements in respect of the product, clearly and concisely*].

I look forward to receiving your letter by [*insert date*] at the latest.

Yours faithfully

函件 57

致当地政府环境管理部门，随函附上平面示意图

Letter 57

To local authority environmental services department, enclosing sketch plans.

尊敬的先生：

我方是负责上述项目的建筑师，我方委托人是［填入委托人姓名］。

随函附上编号为［填入编号］的图纸副本各两份，包括平面示意图和项目总体布局图。如果你方能审查此项目建筑方案，并告知你方接受图中所示的废物收集流程，或者告知你方对此的意见，我方将不胜感激。

你忠诚的

Dear Sir

I act as architect for the above development. My client is [*insert name*].

I enclose two copies of each of drawings numbers [*insert numbers*] showing sketch plans and layout of the proposals and I should be grateful if you would examine the proposals and let me have your acceptance or comments on the refuse collection arrangements indicated.

Yours faithfully

函件 58

致消防管理部门官员，随函附上平面示意图

Letter 58

To fire prevention officer, enclosing sketch plans

尊敬的先生：

我方是负责上述项目的建筑师，我方委托人是［填入委托人姓名］。

随函附上编号为［填入编号］的工程图纸副本各两份，包括平面示意图和此项目的立面图与剖面图。如果你方能审查此建筑方案，并告知你方对此的意见和指示，我方将不胜感激。

［可以加上：］

在考虑你方的意见后，我方将打电话召集设计团队成员会议，以解决难点问题。

你忠诚的

Dear Sir

I act as architect for the above development. My client is [*insert name*].

I enclose two copies of each of drawings numbers [*insert numbers*] showing sketch plans, elevations and sections of the proposal and I should be grateful if you would examine the proposals and let me have your recommendations and observations.

[*You may wish to add*:]

After studying your comments, I will telephone to arrange a meeting with other members of the design team to iron out any difficulties.

Yours faithfully

函件 59

致电话服务供应商，随函附上布局示意图

Letter 59

To telephone service supplier, enclosing sketch layout

尊敬的先生：

谨提及［填入之前与电话服务供应商所有往来信函的日期］的往来信函。随函附上前期场地布局示意图，其中绿色所示为电话服务设施/所有电话设施［视情况取舍］的接入口位置。此为编号为［填入编号］的工程图纸。

［视情况，加上：］

红色所示为需要迁移的电话线缆。

［然后：］

工地在项目进行过程中需要临时的电话服务。在指定主承包商后，主承包商会向你方发出订单，并支付费用。相关的预计工程进度日期为：

工程开工：［填入日期］

工程竣工：［填入日期］

所需电话服务：［填入日期］

需要进行的设施迁移：［视情况填入日期］

如果你方能寄回一份平面图副本，标出电话线路及其他必需设施，［如果需要迁移线缆，加上：］包括你方迁移红色所示的线缆和进行必要整改的费用的确实报价，我方将不胜感激。

你忠诚的

Dear Sir

I refer to previous correspondence of the [*insert dates of all previous letters to and from telephone service provider*]. I enclose two copies of my preliminary layout of the site including the inlet positions of all telephone services/the telephone service [*delete as appropriate*] coloured green. This is drawing number [*insertnumber*].

[*Add, if appropriate:*]

The cable to be diverted is shown coloured red.

[*Then:*]

A temporary service will be required on site for the duration of the project. It will be paid for by the main contractor who will send his order when appointed. The relevant programme dates are as follows:

Commencement of project: [*insert date*].
Completion of project: [*insert date*].
Telephone service required: [*insert date*].
Diversion to be completed: [*insert date if appropriate*].

I should be pleased if you would return one copy of the plan showing the service line [*s*] and any other provision required [*if cable is to be diverted, add:*] including your firm price quotation for carrying out the diversion of the cable coloured red and making good.

Yours faithfully

函件 60

致电力供应商，随函附上布局示意图

Letter 60

To electricity supplier, enclosing sketch layout

尊敬的先生：

谨提及［填入之前与电力供应商所有往来信函的日期］的往来信函。随函附上前期场地布局示意图，其中绿色所示为电力服务设施/所有电力设施［视情况取舍］的接入口位置。此为编号为［填入编号］的工程图纸。

［视情况，加上：］

红色所示为需要迁移的电力干线。

［然后：］

所需的服务为［填入细节］。相关的预计工程进度日期为：

工程开工：［填入日期］

工程竣工：［填入日期］

所需电力服务：［填入日期］

需要进行的设施迁移：［视情况填入日期］

如果你方能寄回一份平面图副本，标出电力线路及其他必需设施，并对工程报出实价，［如果需要迁移电力干线，加上：］包括你方迁移红色所示的线缆和进行必要整改的费用报价。

你忠诚的

Dear Sir

I refer to previous correspondence of the [*insert dates of all previous letters to and from the electricity supplier*]. I enclose two copies of my preliminary layout of the site including the inlet positions of all electrical services/the electricity service [*delete as appropriate*] coloured green. This is drawing number [*insert number*].

[*Add*, *if appropriate*:]

The main to be diverted is shown coloured red.

[*Then*:]

The service [*s*] required is/are [*delete as appropriate then insert details*]. The relevant programme dates are anticipated to be as follows:

Commencement of project: [*insert date*].
Completion of project: [*insert date*].
Electricity service required: [*insert date*].
Diversion to be completed: [*insert date if appropriate*].

I should be pleased if you would return one copy of the plan showing the service line [*s*] and any other provision required together with your firm price quotation for carrying out the work [*if main is to be diverted*, *add*:] including diversion of the main and all necessary making good.

Yours faithfully

函件 61

致燃气供应商，随函附上布局示意图

Letter 61

To gas supplier, enclosing sketch layout

尊敬的先生：

谨提及［填入之前与燃气供应商所有往来信函的日期］的往来信函。随函附上前期场地布局示意图，其中绿色所示为燃气服务设施/所有燃气设施［视情况取舍］的接入口位置。此为编号为［填入编号］的工程图纸。

［视情况，加上：］

红色所示为需要迁移的燃气管道。

［然后：］

相关的预计工程进度日期为：
工程开工：［填入日期］
工程竣工：［填入日期］
所需燃气服务：［填入日期］
需要进行的设施迁移：［视情况填入日期］
如果你方能寄回一份平面图副本，标出燃气管道线路，并对工程报出实价，［如果需要迁移燃气管道，加上：］包括你方迁移红色所示的管道和进行必要整改的费用报价。

你忠诚的

Dear Sir

I refer to previous correspondence of the [*insert dates of all previous letters to and from the gas supplier*]. I enclose two copies of my preliminary layout of the site including the inlet positions of all gas services/the gas service [*delete as appropriate*] coloured green. This is drawing number [*insert number*].

[*Add*, *if appropriate*:]

The main to be diverted is shown coloured red.

[*Then*:]

The relevant programme dates are anticipated to be as follows:

Commencement of project: [*insert date*].
Completion of project: [*insert date*].
Gas service required: [*insert date*].
Diversion to be completed: [*insert date if appropriate*].

I should be pleased if you would return one copy of the plan showing the service line [*s*] together with your firm price quotation for carrying out the work [*if main is to be diverted*, *add*:] including diversion of the main and all necessary making good.

Yours faithfully

函件 62

致自来水供应商，随函附上布局示意图

Letter 62

To water supplier, enclosing sketch layout

尊敬的先生：

谨提及［填入之前与自来水供应商所有往来信函的日期］的往来信函。随函附上前期场地布局示意图，其中绿色所示为所有自来水设施/自来水服务设施［视情况取舍］的接入口位置。此为编号为［填入编号］的工程图纸。

［视情况，加上：］

红色所示为需要迁移的自来水管道。

［然后：］

相关的预计工程进度日期为：
工程开工：［填入日期］
工程竣工：［填入日期］
所需自来水服务：［填入日期］
需要进行的设施迁移：［视情况填入日期］
如果你方能寄回一份平面图副本，标出自来水管道线路，并对工程报出实价，［如果需要迁移自来水管道，加上：］包括你方迁移红色所示的管道和进行必要整改费用报价。

你忠诚的

Dear Sir

I refer to previous correspondence of the [*insert dates of all previous letters to and from the water supplier*]. I enclose two copies of my preliminary layout of the site including the inlet positions of all water services/the water service [*delete as appropriate*] coloured green. This is drawing number [*insert number*].

[*Add*, *if appropriate*:]

The main to be diverted is shown coloured red.

[*Then*:]

The relevant programme dates are anticipated to be as follows:

Commencement of project: [*insert date*].
Completion of project: [*insert date*].
Water service required: [*insert date*].
Diversion to be completed: [*insert date if appropriate*].

I should be pleased if you would return one copy of the plan showing the service line [*s*] together with your firm price quotation for carrying out the work [*if main is to be diverted*, *add*:] including diversion of the main and all necessary making good.

Yours faithfully

函件 63

致公路管理部门，随函附上布局示意图

Letter 63

To highway authority, enclosing sketch layout

尊敬的先生：

谨提及［填入之前与公路管理部门所有往来信函的日期］的往来信函。随函附上前期场地布局示意图，其中标出了计划建设的道路、人行道及出入方式。此为编号为［填入编号］工程的图纸。

我方欣然希望能够获得你方的批准或详细意见，这样就可以继续完成我方的设计。特别是对于以下事项，能否请你方批示：

1. 道路和人行道宽度。
2. 视线。
3. 现有公路的迁移。
4. 现有公路的停用。
5. 道路照明。
6. 获得你方正式批准所申报图纸的程序。
7. 获得你方批准采用图中绿色所示的道路及人行道的程序。

你忠诚的

Dear Sir

I refer to previous correspondence of the [*insert dates of all previous letters to and from the highway authority*]. I enclose two copies of my preliminary layout of the site including all proposed roads, footpaths and means of ingress and egress. This is drawing number [*insert number*].

I should be pleased to receive your approval or comments in detail so that I can make progress in completing my design. In particular, may I have your observations on the following:

[*Delete as appropriate from the following*:]

1. Road and footpath widths.
2. Sight lines.
3. Diversion of existing highway.
4. Stopping of existing highway.
5. Street lighting.
6. Procedure for obtaining formal consent from your authority to the enclosed proposals.
7. Procedure for obtaining consent to the adoption of the roads and footpaths coloured green.

Yours faithfully

函件 64

关于《空气清洁法案》，致环境卫生部门

Letter 64

To environmental health authority, regarding Clean Air Acts

尊敬的先生：

我方明确，上述项目需要符合《空气清洁法案》的特殊设计要求。

现随函附上编号为［填入编号］的工程图纸，希望你方能给予意见。

鉴于设计工作正在进行，我方将于几天内电话联系你方，听取你方的意见。

你忠诚的

Dear Sir

I understand that the above development may be subject to special design requirements under the Clean Air Acts.

I enclose two copies of each of my drawings numbers [*insert numbers*] and I should be pleased to receive your comments.

Since design development is continuing, I will telephone you during the next few days in order to obtain your initial reactions to the scheme.

Yours faithfully

函件 65

致委托人，告知补充要求将使建筑造价高于预算

Letter 65

To client, advising that additional requirements will cause the cost to rise above budget

尊敬的先生：

谨提及我们于昨日［或说明会议日期］的会议。会上讨论了你方的补充要求：

［简要列出补充要求］

我方现广泛地考虑了这些要求对财务方面的影响。如果需要改变设计来满足这些要求，其费用将会大大超出此项目的预算。

因为我方不具有建筑造价核算方面的专业知识，我方已经把细节告知你方的工料测师，他能够更为准确地告知你方可能的超支费用金额。现已获悉工料测量师将于［填入日期］之前做出造价报告，我方期望到时与你方讨论此事。

你忠诚的

抄送：工料测量师

Dear Sir

I refer to our meeting yesterday [*or state the date of the meeting*] to discuss your further requirements. These were:

[*Briefly list the additional requirements*]

I have now had the opportunity to broadly consider the financial effect of these requirements and it appears that, if the design is to be amended to incorporate them, your budget for the project will be exceeded by a substantial amount.

Since I do not have any particular expertise in the field of construction costs, I have given details to your quantity surveyor so that he can advise you about the likely additional cost with more precision. I understand that he will be in a position to report on costs by [*insert date*] and I look forward to discussing the matter with you then.

Yours faithfully

Copy: Quantity surveyor

函件 66

致委托人，随函附上《雇主要求》的最终草稿

此信函仅适用于 WCD98 合同

Letter 66

To client, enclosing final draft of Employer's Requirements

This letter is only suitable for use with WCD 98

尊敬的先生：

我方现已经完成此项目的《雇主要求》的最终草稿，见随函附件。

此文件较长，有两方面的目的。第一，它为承包商准备起草投标书设定了参数。第二，它明确说明了必需的各项要素。请你方仔细阅读，以确认其中涵盖了所有你方重点关心的事项。

在此阶段，希望此文件不会有任何大的改动。为了使工程符合进度计划，如果你方能于［填入日期］之前给予意见，我方将不胜感激。

你忠诚的

Dear Sir

I have now completed and enclose the final draft of the 'Employer's Requirements' for this project.

The document is lengthy. Its aim is twofold. Firstly, to set the parameters within which the contractors can prepare their tenders. Secondly, to make clear those elements which are mandatory. Please read it through carefully to see that I have included all those matters about which you are particularly concerned.

Hopefully, at this stage, any changes to the document will be minor. If the programme is to be met, I should be grateful for your comments by [*insert date*] at the latest.

Yours faithfully

函件 67

致委托人，随函附上详细建筑方案报告

Letter 67

To client, enclosing the detailed proposals report

尊敬的先生：

现欣然附上我方详细建筑方案报告以及编号为［填入编号］的工程图纸，图中展示了该项目详细的建筑方案。

谨确认我方将与你方于［填入日期］［填入时间］在［填入地点］进行面谈。期待与你方讨论建筑方案报告中的各项事宜，并听取你方关于下一阶段工作的指示。

［视情况，加上：］

如果你方能在面谈之前考虑［视情况填入事项］，我方将不胜感激。

你忠诚的

Dear Sir

I have pleasure in enclosing my detailed proposals report and drawings numbers [*insert numbers*] showing my detailed proposals for this project.

I confirm that I will meet you on [*insert date*] at [*insert time*] at [*insert place*] and I look forward to discussing matters arising from my report and taking your instructions for the next stage.

[*If appropriate, add:*]

I should be grateful if you would carefully consider [*insert as appropriate*] before our meeting.

Yours faithfully

函件 68

关于设计要点修正，致委托人

Letter 68

To client, regarding modification of the brief

尊敬的先生：

谨确认你方同意编号为［填入编号］的图纸所示的详细建筑方案需进行如下方面的修正。

［简要描述修订的内容］

我方将尽快完成最终建筑方案。此后的任何对于设计要点的修订将严重影响项目进度。由于必需重新设计，我方将另外收取费用。

你忠诚的

Dear Sir

I confirm your approval of the detailed proposals shown on my drawing number [*insert drawing number*] subject to the following modifications:

[*Briefly describe the modifications*]

I am now completing the final proposals with all speed. Any modifications to the brief after this point would have serious effect on the time schedule for the project and, since redesigning would be necessary, extra fees would be chargeable.

Yours faithfully

函件 69

致委托人，确认进行额外工作的指示

此函仅适用于 SFA/99、CE/99 和 SW/99 合同

Letter 69

To client, confirming instruction to do extra work

This letter is suitable for use with SFA/99, CE/99 and SW/99

尊敬的先生：

谨提及我们于昨日的会议/电话谈话［视情况取舍］。你方指示我方进行［填入指示的简要描述］。

此项工作是在我方雇佣条款所规定的服务范围之外，属于第 5.6 条［SW99 合同时，替换为“第 2 条”］规定的额外工作类。我方将欣然执行此项工作，但我方必须说明，将按时间收取费用，收费标准为［填入雇佣条款中说明的收费率］。

你忠诚的

Dear Sir

I refer to the meeting/telephone conversation [*delete as appropriate*] yesterday when you instructed me to [*insert a brief description of the instruction*].

This work is not included in the services I undertook to perform under the terms of engagement. It falls under the category of extra work and clause 5.6 [*substitute '22' when using SW/99*] refers. I am delighted to carry it out, but I must advise you that I will charge on a time basis at the rate of [*insert the rate stated in the terms of engagement*].

Yours faithfully

第五章　最终建筑方案

尽管在此阶段建筑师的设计工作还没有板上钉钉，但你最后必须强调，建筑方案的修改需要额外支付大笔的费用。

函件 70

关于规模、外形、布局或造价的修正，致委托人

Letter 70

To client, regarding modifications to size, shape, location or cost

尊敬的先生：

谨确认此项目详细的设计工作已经完成，我方正打算着手准备与施工相关的必要的生产信息。

我方相信你方清楚现在改变主意会引起的问题。任何对于规模、外形、布局或造价方面的更改会造成已经完成的大量工作的夭折，而且会造成工期延误和产生额外的费用。

你忠诚的

Dear Sir

I confirm that the detailed design work for the above project is now complete and I am about to commence preparing the production information necessary for construction.

I know that you appreciate the problems caused by changes of mind at this stage. Any further alterations to the size, shape, location or cost of the project will make much of the completed work abortive which, in turn, will involve delays and additional fees.

Yours faithfully

函件 71

关于迅速作出决定，致委托人

Letter 71

To client, regarding promptness of decisions

尊敬的先生：

我方认为此时需提醒你方应该检查一下关于迅速作出决定的安排。

我方相信你方已经清楚在合同执行过程中，尽管我方可以在合适的时候给你方提供专业建议，但很多问题还是一定要请你方亲自作决定。

在这种情况下，你方必须能够马上做出决定，以避免工程延期或产生额外的费用。

你方可以指定专人全权负责，在紧急情况下迅速做出令人满意的决定。

你忠诚的

Dear Sir

This is perhaps an opportune moment to ask you to review your arrangements for providing urgent decisions.

I am sure you appreciate that there will be certain points during the currency of the contract which must be referred to you, although I am always ready to give you my professional advice where appropriate.

In such circumstances, it is essential to obtain a quick decision in the interests of avoiding delays and extra costs.

If it is possible for you to nominate one person with full powers to make any urgent decisions that would be entirely satisfactory.

Yours faithfully

第六章　生产信息

多数建筑师都认为这是最忙的一个阶段。在这一阶段要做好所有的施工图纸和表单，取得各种重要的批准许可及着手安排分包商、供应商和主要合同工程招投标事宜。如果项目使用的是 WCD98 合同，建筑师则忙于整理完成《雇主要求》。《雇主要求》实质上是一个绩效考量的技术规范，因而它的编写是至关重要的。

幸运的是，大量的行政信函都可以使用标准信函格式。一般情况下，建筑师可以按常规方式通过《建筑规范》的审查来申请工程的批准许可，这也就是将来的工作模式。如果你想依据其他的方式通过《建筑规范》这一关，你就必须另行拟写合适的信函。

如果从收集投标人名单算起的话，投标可以说是一个很长的过程。关键是要尽早着手进行。建筑师都会有一些有着良好施工纪录的承包商名单，但建筑师必须与委托人再次检查。不要轻易代表委托人接受承包商投标。当一家公司要求被列入投标人名单时，如果你对其经济地位或实力有任何疑问，一定要进行核查或立即告知你的委托人。不要向委托人推荐承包商。这样做是不合适的，而且也是很危险的。因为你可能不清楚这家公司的人事变动或经济状况的转变，而这些变化可能会导致与你对他们的期望完全不同的结果。

函件 72

致委托人，索要申请《建筑规范》批准的费用

Letter 72

To client, requesting fees for Building Regulations application

尊敬的先生：

我方期望能够于［填入日期］申请《建筑规范》批准。当地部门须按收费明细表收取费用。当地部门只有在收到足额的费用后才会处理申请事宜。

我方已经计算了此项申请必需支付的费用，金额为［填入金额］。我方希望能收到你方按此金额开出的以［填入当地部门的名称］为收款人的支票。请注意，当地部门依据《建筑规范》完成第一次工地检查后，你方须按收费标准支付下一笔费用。

如果需要更多的信息，请告知我方。

你忠诚的

Dear Sir

I anticipate being in a position to make application for Building Regulation approval on the [*insert date*]. The local authority is required to charge fees in accordance with a schedule of charges. The application will not be considered until the correct fee is received by the authority.

I have calculated the fee payable in respect of this application as [*insert amount*] and I should be pleased to receive your cheque for this amount which you should draw in favour of [*insert name of authority*]. Please note that after the authority have carried out their first inspection on site, a further fee will become payable in accordance with the scale.

Please let me know if you require further information.

Yours faithfully

函件 73

致委托人，索要申请规划许可和《建筑规范》批准的费用

Letter 73

To client, requesting fees for Planning and Building Regulations application

尊敬的先生：

我方期望能够于［填入日期］申请完全规划许可和《建筑规范》的批准。每项申请都要支付费用。当地管理部门只有在收到足额的费用后才会处理申请事宜。

我方已经计算了此项申请必需支付的费用如下：
规划许可申请：［填入金额］
《建筑规范》批准申请：［填入金额］

我方十分希望你方能开出一张金额为［填入总金额］的以［填入当地管理部门名称］为受款人的支票。如果你方将此支票寄给我方，我方将附于申请书提交。请注意，当地管理部门按《建筑规范》规定，完成第一次场地检查后，你方须按收费标准支付下一笔费用。

如果需要更多的信息，请告知我方。

你忠诚的

Dear Sir

I anticipate being in a position to make application for full planning permission and approval under the Building Regulations on the [*insert date*]. A separate fee is payable in respect of each application. Until the fee is received by the authority, the application will not be considered.

I have calculated the fees payable as follows:

Planning application: [*insert amount*].
Building Regulations application: [*insert amount*].

I should be pleased if you would draw a single cheque in the sum of [*insert total amount*] in favour of [*insert name of authority*]. If you will send the cheque to me, I will attach it to the application. Please note that after the authority have carried out their first inspection on site under the Building Regulations, a further fee will become payable in accordance with the scale.

Please let me know if you require any further information.

Yours faithfully

函件 74
致规划管理部门，要求批准保留事项

Letter 74
To planning authority, requesting approval of reserved matters

尊敬的先生：

谨提及［填入日期］签发的关于上述项目的［填入编号］初步规划许可。我方现代表我方的委托人［填入姓名］正式申请批准以下保留事项：

［按照原许可内容列出保留事项。如果只申请批准某几项事宜，则只需列出这几项。］

为支持我方申请，随函附上编号为［填入编号］的工程图纸3份。
如果申请批准过程中出现任何困难，请电话通知我方，以便尽快解决，对此我方将不胜感激。

你忠诚的

Dear Sir

I refer to outline planning permission number [*insert number*] granted on the [*insert date*] in respect of the above development. I now formally apply on behalf of my client [*insert name*] for approval of the following reserved matters:

[*List the reserved matters exactly as they appear on the original permission. If approval is sought for certain matters only, list only those matters.*]

In support of this application, I enclose three copies of drawings numbers [*insert numbers*].

If there is likely to be any difficulty in obtaining approval, I should appreciate a telephone call so that such difficulties can be resolved as soon as possible.

Yours faithfully

Copy: Client

函件 75

致规划管理部门，申请完全规划许可

Letter 75

To planning authority, applying for full planning approval

尊敬的先生：

谨提及我方与你方［填入姓名］已经就此规划和［填入日期］签发的初步规划许可进行了讨论。现谨代表我方委托人正式申请完全规划许可。

随函附上下列材料：

1. 完成的申请表 3 份。
2. 编号为［填入编号］的工程图纸 3 份。
3. 费用［视情况填入金额。如果不需支付费用，删去此条］。
4. 证明书 A/B/C/D［视情况取舍］。

现证实随函附上的公告曾经于［填入日期］在［视情况填入地方报纸名称］上发布，并且在工地内显眼位置展示了 7 天［填入起止日期］。

如果出现任何可能妨碍或延误发放完全规划许可的事情，请电话通知我方，以便尽快解决。对此我方将不胜感激。

你忠诚的

Dear Sir

I refer to my discussions with your [*insert name*] with regard to this proposal and outline planning permission dated [*insert date*]. I now formally submit an application for full planning permission on behalf of my client [*insert name*].

I enclose:

1. Three copies of the completed form of application.
2. Three copies of drawings numbers [*insert numbers*].
3. The fee of [*insert amount of fee as appropriate. If no fee is payable, delete this item*].
4. Certificate A/B/C/D [*delete as appropriate*].

[*Add, if appropriate:*]

I hereby certify that the enclosed advertisement has been published on the [*insert dates*] in the [*insert name of local newspaper and add, if appropriate:*] and the said advertisement has been displayed in a prominent position on the site for a period of 7 days from [*insert date*].

If there are any matters which could prevent or delay the granting of full planning permission, I should appreciate a telephone call so that such matters can be resolved as quickly as possible.

Yours faithfully

函件 76

致规划管理部门，申请完全规划许可和《建筑规范》批准

Letter 76

To planning authority, applying for full planning permission and Building Regulations approval

尊敬的先生：

我方已经与你方规划处的［填入姓名］和建筑物控制处的［填入姓名］分别进行了多次讨论。初步规划许可［填入号码］已经于［填入日期］发放。现谨代表我方的委托人向你方申请完全规划许可及《建筑规范》批准。

随函附上下列材料：

1. 完成的申请表5份。
2. 编号为［填入编号］的工程图纸5份。
3. 费用［视情况填入金额。如果不需支付费用，删去此条］。
4. 证明书A/B/C/D［视情况取舍］。

［视情况，加上：］

现证实随函附上的公告曾经于［填入日期］在［视情况填入地方报纸名称］上发布，并且在工地内显眼位置展示了7天［填入起止日期］。

我方已经与HM生产制造检查员（健康与安全管理人员）进行了正式磋商，并随函附上了相应的证明书。

如果出现任何可能妨碍或延误发放规划许可或《建筑规范》批准的事情，请电话通知我方，以便尽快解决。对此我方将不胜感激。

你忠诚的

抄送：委托人

Dear Sir

I refer to my discussions with [*insert name*] in your planning department and [*insert name*] in building control. Outline planning permission number [*insert number*] was granted on [*insert date*]. I now formally submit an application for full planning permission and approval under the Building Regulations on behalf of my client [*insert name*].

I enclose:

1. Five copies of the completed form of application.
2. Five copies of drawings numbers [*insert numbers*].
3. The fee of [*insert amount of fee as appropriate. If no fee is payable, delete this item*].
4. Certificate A/B/C/D [*delete as appropriate*].

[*Add, if appropriate:*]

I hereby certify that the enclosed advertisement has been published on the [*insert dates*] in the [*insert name of local newspaper and add, if appropriate:*] and the said advertisement has been displayed in a prominent position on the site for a period of 7 days from [*insert date*].

Formal consultations have been carried out with HM Factory Inspectorate (Health and Safety Executive) and the appropriate certificate is enclosed.

If there are any matters which could prevent or delay granting of planning permission or approval under the Building Regulations, I should be grateful if you would telephone me immediately so that such matters can be resolved without delay.

Yours faithfully

Copy: Client

函件 77

致当地政府部门，申请《建筑规范》批准

Letter 77

To local authority, requesting Building Regulations Approval

尊敬的先生：

我方已经与你方建筑物控制处［填入姓名］进行了多次讨论。现谨代表我方的委托人［填入姓名］正式申请《建筑规范》批准。

随函附上下列材料：

1. 完成的申请表 2 份。
2. 编号为［填入编号］工程的图纸 2 份。
3. 费用［视情况填入金额。如果不需支付费用，删去此条］。

［视情况，加上：］

我方已经与消防管理部门进行了正式磋商/已经着手与消防管理部门进行正式磋商［视情况取舍］，随函附上消防证书/消防证书将随后寄给你方［视情况取舍］。

如果出现任何可能妨碍或延误发放规划许可或《建筑规范》批准的事情，请电话通知我方，对此我方将不胜感激。

你忠诚的

Dear Sir

I refer to my discussions with [*insert name*] in your building control department. I now formally submit an application for approval under the Building Regulations on behalf of my client [*insert name*].

I enclose:

1. Two copies of the completed form of application.
2. Two copies of drawings numbers [*insert numbers*].
3. The fee of [*insert amount of fee as appropriate. If no fee is payable, delete this item*].

[*Add, if appropriate:*]

I have commenced/completed [*delete as appropriate*] formal consultations with the fire authority and a fire certificate is enclosed/will be forwarded in the near future [*delete as appropriate*].

I should appreciate a telephone call if any point arises which might prevent or delay approval.

Yours faithfully

函件 78

致消防管理部门，申请消防证书

Letter 78

To fire authority, applying for fire certificate

尊敬的先生：

我方深知鉴于上述项目的性质，我方需要与 HM 生产制造检查员进行正式磋商，以获得《1971 年消防法案》规定的消防证书。

现附上编号为［填入编号］的工程图纸各两份。其中详细展示了我方代表委托人［填入姓名］特此提交的规划图的细节。

《建筑规范》批准申请表将于几天内提交到［填入地址］的［填入部门名称］。如果需要进一步了解任何信息，务请告知，我方将不胜感激。

你忠诚的

抄送：委托人

Dear Sir

I understand that the above development is of a nature that requires me to carry out formal consultations with HM Inspectorate in order to obtain a fire certificate under the Fire Precautions Act 1971.

I enclose two copies of each of drawings numbers [*insert numbers*] showing details of my proposals which I hereby submit on behalf of my client [*insert name*].

Application for approval under the Building Regulations is being submitted to [*insert name of authority*] at [*insert address*] during the course of the next few days and I should appreciate it if you would inform me if you require any further information.

Yours faithfully

Copy: Client

函件 79

关于主承包商投标人名单，致委托人

Letter 79

To client, regarding main contractor tender list

尊敬的先生：

按照此项目议定的时间表，我方将代表你方于［填入日期］前后按《1996 年单一阶段招投标程序规范》邀请投标人。

我方将与你方详细讨论投标程序，但十分希望你方现在能告知是否有意提名某承包商为投标人。我方已经考虑到了几家合适的承包商。在完成必要的问询之后，我方将与你方面议，拟订最终名单。

你忠诚的

Dear Sir

In accordance with the agreed timetable for this project, I shall be inviting tenders on your behalf on or about the [*insert date*] using the Code of Procedure for Single Stage Selective Tendering 1996.

I will discuss the tendering procedure with you in more detail, but at this stage I should be pleased to know if you wish to nominate any contractors for inclusion on the tender list. I already have several contractors in mind who could be suitable, and when I have completed the necessary enquiries I will meet you to draw up the final list.

Yours faithfully

函件 80

关于两段招标情况下的主承包商投标人名单，致委托人

Letter 80

To client, regarding main contractor tender list if two stage tendering is to be used

尊敬的先生：

按照此项目议定的时间表，并根据近期会议的讨论内容，我方将代表你方于［填入日期］前后按《1996 年两阶段招投标程序规范》邀请投标人。

十分希望现在你方能告知是否有意提名某承包商为投标人。我方已经考虑到了几家合适的承包商。在完成必要的问询之后，我方将与你方面谈拟订最终名单。

你忠诚的

Dear Sir

In accordance with the agreed timetable for this project, I shall be inviting tenders on your behalf on or about the [*insert date*] using the Code of Procedure for Two Stage Selective Tendering 1996 as discussed at our recent meeting.

I should be pleased to know if you wish to nominate any contractors for inclusion on the tender list. I already have several contractors in mind who could be suitable, and when I have completed the necessary enquiries I will meet you to draw up the final list.

Yours faithfully

函件 81

致委托人，确认投标细节

Letter 81

To client, confirming tender details

尊敬的先生：

按照我们［填入日期］的讨论内容。我方认为有必要确认你方关于邀请投标人而作出的如下决定：

1. 准备邀请的投标人：［名单］。
2. 投标期限：［填入期限］。
3. 接收标书的日期、时间和地点：［填入细节］。
4. 接收标书的期限：［填入期限］。
5. 采用《1996 年单一阶段招投标程序规范》的备选方案 1/2［视情况取舍］。

你忠诚的

Dear Sir

Following our discussions on the [*insert date*], I thought it would be useful to confirm the decisions you made with regard to the invitation of tenders as follows:

1. Contractors to be invited to tender: [*list*].
2. Period of time allowed for tendering: [*insert period*].
3. Date, time and place for receipt of tenders: [*insert details*].
4. Tender to remain open for acceptance for: [*insert period*].
5. Alternative 1/2 [*delete as appropriate*] of the Code of Procedure for Single Stage Selective Tendering 1996 to apply.

Yours faithfully

函件 82

致礼物赠予者，退还礼物

Letter 82

To donor，returning gift

尊敬的先生：

今天收到你方寄来的礼物，深表感谢。

相信你方明白我方不会允许任何事情对我方的职业操守造成哪怕极微小的影响。因此，我方按照自己一贯的原则，将礼物退还给你方。

不过，非常感谢你方这样做的心意。

你忠诚的

Dear Sir

Thank you for your gift which I received today.

I am sure that you understand that I cannot allow anything to throw the slightest doubt upon my professional integrity. For this reason I am following my usual policy and returning the gift to you.

That is not to say, however, that I do not appreciate the thought which prompted your action.

Yours faithfully

函件 83

致有希望被指定的分包商/分包人，询问是否有意提交标书

Letter 83

To prospective nominated or named sub-contractor/person, enquiring if willing to submit a tender

尊敬的先生：

我方的委托人［填入姓名］指示我方准备一份因对［填入工程内容］感兴趣而有意对此项目投标的公司名单。如果你方希望被列入其中，请于［填入日期］之前书面通知我方。如果你方此次不能对该项目投标，不会影响你方对我方负责的其他项目的投标资格。但请注意，你方同意投标并不能保证你方能够获得投标邀请。下列信息供你方参考：

［按以下条目，填入项目信息：］

1. 项目名称：
2. 雇主姓名：
3. 建筑师姓名：
4. 工程量估算师姓名：
5. 顾问名称：
6. 工地地址：
7. 工程概述：
8. 大概造价范围：
9. 主合同格式：
10. 要删去的合同条款：
11. 特殊附加条款：
12. 分包合同格式：
13. 要删去的合同条款：
14. 特殊附加条款：
15. 合同按非盖印合同/契约合同执行［视情况取舍］。
16. 主合同中预期进驻场地时间：
17. 主合同工程竣工期限：
18. 发出投标文件的大约日期：
19. 投标期限：
20. 接收标书的期限为［填入周数］。
21. 主合同中的预定违约金：

你忠诚的

Dear Sir

I have been instructed by my client, [*insert name*], to prepare a list of firms willing to tender for [*insert nature of work*] on the above project. Please inform me in writing, not later than [*insert date*] if you wish to be included. If you are unable to tender on this occasion, it will not prejudice your inclusion on tender lists for other projects under my direction, but you should note that your agreement to tender does not guarantee that you will receive an invitation to do so. The following is set out for your information:

[*Insert information appropriate to your project in the following items*:]

1. Name of project:
2. Name of employer:
3. Name of architect:
4. Name of quantity surveyor:
5. Name of consultants:
6. Site address:
7. General description of work:
8. Approximate cost range:
9. Form of main contract to be used:
10. Contract clauses to be deleted:
11. Special additional clauses:
12. Form of sub-contract to be used:
13. Contract clauses to be deleted:
14. Special additional clauses:
15. The contract is to be executed under hand/as a deed [*delete as appropriate*].
16. Anticipated date for possession in main contract:
17. Period for completion of the main contract works:
18. Approximate date for despatch of tender documents:
19. Tender period:
20. Tender to remain open for acceptance for [*insert number*] weeks.
21. Liquidated damages in main contract:

Yours faithfully

函件 84

致有希望的承包商，随函附上委托人调查表

Letter 84

To prospective contractor, enclosing questionnaire

尊敬的先生：

我方打算于［填入日期］为以上项目邀请投标人。此项工程包括［填入工程的简要描述］。如果你方希望被列入投标人名单，请在［填入日期］之前告知我方如下方面的信息：

1. 公司全体董事的姓名和地址。
2. 注册办公地点。
3. 公司的股份资产。
4. 过去三年的年产值。
5. 办公室职员的人数及职务。
6. 每个工种永久雇用的工地施工人员人数。
7. 经过培训的工地管理人员人数。
8. 目前合同在建工程的数量和造价。
9. 列出三个近期竣工且与招标项目特点相似的工程的地址、竣工日期和工程造价。
10. 尽可能提供证明的上述第 9 条中注明的项目委托人、建筑师及工料测量师的姓名和地址。

你忠诚的

Dear Sir

I anticipate inviting tenders for the above project on or about [*insert date*]. The Works will consist of [*insert brief description*]. If you wish to be considered for inclusion on the list of contractors invited to tender, please let me have the following information by [*insert date*]:

1. Names and addresses of all directors.
2. Address of your registered office.
3. Share capital of the firm.
4. Annual turnover during the last three years.
5. Number and positions of all office-based staff.
6. Number of site operatives permanently employed in each trade.
7. Number of trained supervisory staff permanently on site.
8. Number and value of current contracts on site.
9. Address, date of completion and value of three projects of similar character to that for which tenders are to be invited and which have been carried out by your firm recently.
10. Names and addresses of clients, architects and quantity surveyors connected with the projects noted in 9 above and to whom reference may be made.

Yours faithfully

函件 85

关于承包商是否合适，致公断人

Letter 85

To referee，regarding suitability of contractor

尊敬的先生：

我方从［填入承包商名称］处得知，他方最近为你方完成了一个项目。我方打算将此承包商列入以上工程的投标人名单。此工程包括［填入工程的简要描述］。

如果你方能书面回答下面的问题，并用附上的粘好邮票的信封寄回，我方将不胜感激。我方会对你方的回答及你方的其他意见保密。

1. 你方会再用这家公司吗？
2. 相对于要求的质量，你怎样评价此公司的工艺质量？
3. 你方对工地管理满意吗？
4. 你方对公司领导层的组织管理满意吗？
5. 其员工对你方有帮助吗？工作效率高吗？
6. 承包商、分包商、指定分包商、供应商及雇主代理人之间的关系是否良好？
7. 据你方所知，承包商对分包商和供应商付款是否及时、足额？
8. 据你方经历，此公司通常能按时完工吗？
9. 你方认为他们是否以良好且理性的态度面对索赔事项？

你忠诚的

Dear Sir

[*Insert name of contractor*] have informed me that they have recently completed a project for you. I am considering including them in the list of tenderers to carry out the above work which will consist of [*insert brief description of the work*].

I should be grateful if you would complete the following questions and return them to me in the enclosed stamped addressed envelope. Your answers and any other comments you care to make will remain strictly confidential.

1. Would you use this firm again?
2. How would you describe the quality of workmanship relative to the quality specified?
3. Was site supervision satisfactory?
4. Was head office organisation satisfactory?
5. Were staff helpful and efficient?
6. Were relations good between the contractor, sub-contractors, nominated sub-contractors, suppliers and employer's licensees?
7. So far as you are aware, did the contractor pay his sub-contractors and suppliers promptly and in full?
8. Does this firm normally complete on time in your experience?
9. Do you consider their attitude to claims fair and reasonable?

Yours faithfully

函件 86

致承包商，询问是否有意提交标书

Letter 86

To contractor, enquiring if contractor is willing to submit a tender

尊敬的先生：

我方的委托人［填入姓名］指示我方准备一份有意对此项目投标的公司名单。如果你方希望被列入其中，请于［填入日期］之前书面通知我方。如果你方此次不能对该项目投标，并不会影响你方对我方负责的其他项目的投标资格。但请注意，你方同意投标并不能保证你方就能够获得投标邀请。

投标程序遵循《1996 年单一阶段招投标程序规范》。我方默认所有希望被列入投标人名单的公司都已经明确此规范的内容。下列信息供你方参考：

［按以下条目，填入项目信息：］

1. 项目名称：
2. 雇主姓名：
3. 建筑师姓名：
4. 工料测量师姓名：
5. 顾问姓名：
6. 项目地址：
7. 工程概述：
8. 大概造价范围：
9. 指定分包商预期被使用条款：
10. 使用的合同格式：
11. 要删去的合同条款：
12. 特殊附加条款：
13. 工程概算表的审查与修正：使用方案 A/B［视情况取舍］。
14. 合同按非盖印合同/契约合同执行［视情况取舍］。
15. 预期进驻场地日期：

（转下页）

Dear Sir

I have been instructed by my client, [*insert name*], to prepare a list of firms willing to tender for the above project. Please inform me in writing, not later than [*insert date*], if you wish to be included. If you are unable to tender on this occasion, it will not prejudice your inclusion on tender lists for other projects under my direction, but you should note that your agreement to tender does not guarantee that you will receive an invitation to do so.

The tendering procedure will be in accordance with the Code of Procedure for Single Stage Selective Tendering 1996 and all firms wishing to be included on the tender list will be deemed to have fully informed themselves of its contents. The following is set out for your information:

[*Insert information appropriate to your project in the following items*:]

1. Name of project:
2. Name of employer:
3. Name of architect:
4. Name of quantity surveyor:
5. Name of consultants:
6. Site address:
7. General description of work:
8. Approximate cost range:
9. Items for which it is anticipated that nominated sub-contractors will be used:
10. Form of contract to be used:
11. Contract clauses to be deleted:
12. Special additional clauses:
13. Examination and correction of priced bills: Alternative 1/Alternative 2 [*delete as appropriate*] will apply.
14. The contract is to be executed under hand/as a deed [*delete as appropriate*].
15. Anticipated date for possession:

[*continued*]

函件 86 续表

Letter 86 continued

16. 工程竣工期限：
17. 大约的发出投标文件日期：
18. 投标期限：
19. 接收标书的期限［填入周数］。
20. 预定违约赔偿金：
21. 要求的保证金：

你忠诚的

16. Period for completion of the Works:
17. Approximate date for despatch of tender documents:
18. Tender period:
19. Tender to remain open for acceptance for [*insert number*] weeks.
20. Liquidated damages:
21. Bond required:

Yours faithfully

函件 87

致未中标的分包商和供应商

Letter 87

To unsuccessful sub-contractors and suppliers

尊敬的先生：

你方对［填入对工程或货物的描述］的投标未中。

我方收到的标书如下：

［按渐增顺序列出标书金额。只要没有使投标人与金额相对应的可能性，就可以按字母顺序列出投标人名称］

请相信，我方将很乐意利用以后的机会与你方再次联系。

你忠诚的

Dear Sir

Your tender for [*insert description of work or goods*] was not successful.

Tenders received were as follows:

[*List amounts in ascending order. Names of tenderers may also be listed, separately in alphabetical order unless to do so would make it possible to link tenderers with amounts.*]

You may be assured that I will be happy to approach you again on future occasions.

Yours faithfully

函件 88

关于致指定或提名分包商和供应商的合同意向书，致委托人

Letter 88

To client, regarding letter of intent to nominated or named sub-contractors and suppliers

尊敬的先生：

根据你方于［填入日期］的指示，现随函附上合同意向书草稿，并建议你方寄给以下分包商/供应商［视情况取舍］。

［列出公司名称以及当前的工程或需要的货物］

欣然希望尽快收到你方的协议。

你忠诚的

Dear Sir

In response to your instructions received on the [*insert date*], I enclose a draft letter of intent which I propose to send to the following sub-contractors/suppliers [*delete as appropriate*]:

[*List firms and immediate work or goods required*]

I should be pleased to receive your agreement as soon as possible.

Yours faithfully

函件 89

致提名或指定的分包商或供应商：合同意向书

Letter 89

To nominated or named sub-contractor or supplier：letter of intent

尊敬的先生：

按我方委托人［填入姓名］的指示，现通知你方，你方在［填入日期］对［填入工程或货物］的金额为［填入金额］的投标可以接受。我方将告知主承包商在签订主合同后，与你方签订分包合同/向你方发出订单［视情况取舍］。

我方的委托人无意将此函作为有约束力的合同的证明。不过，我方的委托人准备指示你方进行［填入有限的工程或货物细节］。无论因为何种原因，如果对此工程/货物［视情况取舍］没有与你方签订合同，我方委托人的责任就只限于通过主承包商对［填入有限的工程或货物的描述］付款。

如果没有书面订单，你方不需要进行/提供标书中的其他工程/供应标书中的其他货物［视情况取舍］。在任何情况下，我方的委托人和我方都不承担任何其他责任。

你忠诚的

Dear Sir

My client, [*insert name*], has instructed me to inform you that your tender of the [*insert date*] in the sum of [*insert amount*] for [*insert the nature of the work or goods*] is acceptable and that I intend to instruct the main contractor to enter into a sub-contract/place an order [*delete as appropriate*] after the main contract has been signed.

It is not my client's intention that this letter should be evidence of a binding contract. However, my client is prepared to instruct you to [*insert details of the limited nature of the work or goods required*]. If, for any reason whatsoever, no contract is entered into with you for this work/these goods [*delete as appropriate*], my client's commitment will be strictly limited to payment, through the main contractor, for [*insert the limited nature of the work or goods required*].

No other work/goods [*delete as appropriate*] included in your tender must be carried out/supplied [*delete as appropriate*] without a further written order. No further obligation is placed upon my client and no obligation whatsoever, under any circumstances, is placed upon me.

Yours faithfully

函件 90

致委托人，确认预定违约赔偿金额

Letter 90

To client, confirming the amount of liquidated damages

尊敬的先生：

根据我们［填入日期］的会议，我方现确认你方同意填入合同文件的预定违约赔偿金额为每周［填入金额］。此金额的计算方式如下：

［列出涉及的各项周金额和合计金额］

你忠诚的

抄送：工料测量师

Dear Sir

Following our meeting of the [*insert date*], I confirm that the figure which you have agreed as liquidated and ascertained damages to be inserted in the contract documents is £ [*insert amount*] per week. The figure is calculated as follows:

[*List and total the weekly sums taken into account*]

Yours faithfully

Copy: Quantity surveyor

第七章　工程量清单

这一阶段需要用到的标准信函很少。你将解决上一阶段悬而未决的各类事项，解答工料测量师的问题，并且还有可能完成部分图纸。

你必须决定是否推荐委任一名工程监督员来负责制定工程量清单中的适当条款。有些委托人开始会很抵制雇佣专员，因为他们认为这项工作本来是你的职责，反对为此支付额外的费用。为此你必须坚持要求。

在这一阶段，你可能会意识到有可能会推迟发出标书。一旦确定了准备推迟多久，你就要通知投标人。这样做不仅是出于礼貌，而是完全有必要让投标人知情。否则，你可能会收到被退回的文件，这会给你造成尴尬。

函件 91

如果需要工程监督员，致委托人

Letter 91

To client, if clerk of works required

尊敬的先生：

现在你方应该考虑任命一名工程监督员。此工程监督员的职责是代表你方并在我方的指挥下，对工程进行检查。

我方认为鉴于此工程的规模/复杂性/性质［视情况取舍］，需要不断地/经常地［视情况而定］检查。按照雇佣条款的第 2.5 条［使用 SW/99 合同时，去掉这一部分］，我方建议你方雇佣一名全职/兼职［视情况取舍］工程监督员。

按惯例，你方可直接指定工程监督员，但是我方很乐意就行政任用细节方面给你方提供建议。希望你方考虑此事后电话联系我方。

你忠诚的

Dear Sir

This appears to be a suitable time to consider the appointment of a clerk of works for this project. The duty of a clerk of works is to act as an inspector of the works on your behalf and under my direction.

It is my view that the size/complexity/nature [*delete as appropriate*] of the work demands constant/frequent [*delete as appropriate*] inspections and I recommend the employment of a clerk of works on a full-time/part-time [*delete as appropriate*] basis in accordance with clause 2.5 [*delete the remainder of the sentence after 'basis' when using SW/99*] of the terms of engagement.

It is normal practice for you to appoint the clerk of works directly, but I shall be happy to advise you about the administrative details involved. Perhaps you will telephone when you have had a chance to consider the matter.

Yours faithfully

函件 92

如果投标日期推迟，致投标人名单上的所有承包商

Letter 92

To all contractors on tender list, if date delayed

尊敬的先生：

谨提及你方［填入日期］的来函，信中提到你方有意对以上项目提交标书。

由于出现了不可预见的情况，我方不得不重新估计发出投标文件的日期。目前预计的投标日期为［填入日期］。

如果你方能够检查自己的计划安排并尽快发函确认你方仍愿意提交标书，我方将不胜感激。

你忠诚的

Dear Sir

I refer to your letter of the [*insert date*] indicating your willingness to submit a tender for the above project.

Unforeseen circumstances have obliged me to reassess the date for despatch of tender documents. The date for despatch is now expected to be [*insert date*].

I should be grateful if you would check your programme and confirm, by return if possible, that you are still willing to submit a tender.

Yours faithfully

第八章　投标活动

建议你方遵循《1996年单一阶段招投标程序规范》中规定的程序。［也可选用《1996年两段招投标程序规范》和《1996年设计施工招投标程序规范》］标准信函就是按此规范写成的。及时向你的委托人汇报情况。通常委托人希望开启标书时自己能在场。如果他不在场，最好在有证人在场的情况下开启标书。

函件 93

致承包商，邀请承包商投标，附工程量清单

此信函不适用于 WCD98 合同

Letter 93

To contractor, inviting him to tender if bills of quantities included

This letter is not suitable for use with WCD 98

尊敬的先生：

谨提及你方于［填入日期］的来信，信中表示有意对上述项目投标，现欣然附上下列文件：

1. 工程量清单两份。
2. 编号为［填入编号］的工程图纸各两份，其展示了工程的规模和特点，这些图纸将成为合同图纸。
3. 投标表格两份。
4. 填好地址的信封一个，以便寄回标书及与之相关的说明材料。

请注意以下事项：

a. 请在［填入地点］查看所有完整的工程图纸。

b. 请与［填入姓名、电话］联系安排工地调查。

c. 投标将遵循《1996 年单一阶段招投标程序规范》进行。

d. 工程概算表的审查和调整——将使用规范第 6 条的方案 1/2［视情况取舍］。

请将填好的投标表格装入附上的信封中封好，在［填入日期］［填入时间］之前寄到［填入地方］。

安全收到此函及附件后请告知我方，并请确认你方将按此指示提交标书。

你忠诚的

抄送：雇主

工料测量师

Dear Sir

I refer you to your letter of the [*insert date*] in which you expressed a willingness to submit a tender for the above project. I now have pleasure in enclosing the following:

1. Two copies of the bills of quantities.
2. Two copies of each of drawings numbers [*insert numbers*] giving a general indication of the scope and character of the Works. These will become the contract drawings.
3. Two copies of the form of tender.
4. An addressed envelope for the return of the tender and instructions relating thereto.

Please note the following:

a. A full set of drawings may be inspected at [*insert place*].
b. The site may be inspected by arrangement with [*insert the person and telephone number*].
c. Tendering will be in accordance with the Code of Procedure for Single Stage Selective Tendering 1996.
d. Examination and adjustment of priced bills-Alternative 1/Alternative 2 [*delete as appropriate*] of clause 6 of the Code will apply.

The completed form of tender is to be sealed in the endorsed envelope provided and must arrive at [*insert place*] not later than [*insert time*] on [*insert date*].

Please acknowledge safe receipt of this letter together with the enclosures noted and confirm that you will submit a tender in accordance with these instructions.

Yours faithfully

Copies: Employer
Quantity surveyor

函件 94

致承包商，邀请投标，不附上工程量清单

此函件不适用于 WCD98 合同

Letter 94

To contractor, inviting him to tender if bills of quantities not included

This letter is not suitable for use with WCD 98

尊敬的先生：

谨提及你方于［填入日期］的来信，信中表示有意对上述项目投标，现欣然附上下列文件：

1. 工程详细说明/费率明细表/活动明细表/工程明细表［视情况取舍］两份。
2. 编号为［填入编号］的工程图纸各两份，这些将成为合同图纸。
3. 投标表格两份。
4. 填好地址的信封一个，以便寄回标书及与之相关的说明材料。

请注意以下事项：

a. 请与［填入姓名、电话］联系安排工地调查。
b. 投标将遵循《1996 年单一阶段招投标程序规范》进行。

请将填好的投标表格装入附上的信封中封好，在［填入日期］［填入时间］之前寄到［填入地方］。

安全收到此函及附件后请告知我方，并请确认你方将按此指示提交标书。

你忠诚的

抄送：雇主
　　　工料测量师

Dear Sir

I refer to your letter of the [*insert date*] in which you expressed willingness to submit a tender for the above project. I now have pleasure in enclosing the following:

1. Two copies of the specification/schedule of rates/schedule of activities/schedule of work [*delete as appropriate*].
2. Two copies of each of drawings numbers [*insert numbers*]. These will become the contract drawings.
3. Two copies of the form of tender.
4. An addressed envelope for the return of the tender and instructions relating thereto.

Please note the following:

a. The site may be inspected by arrangement with [*insert the person and telephone number*].
b. Tendering will be in accordance with the Code of Procedure for Single Stage Selective Tendering 1996.

The completed form of tender is to be sealed in the endorsed envelope provided and must arrive at [*insert place*] not later than [*insert time*] on [*insert date*].

Please acknowledge safe receipt of this letter together with the enclosures noted and confirm that you will submit a tender in accordance with these instructions.

Yours faithfully

Copies: Employer
Quantity surveyor

函件 95

致承包商，邀请投标

此函件仅适用于 WCD98 合同

Letter 95

To contractor, inviting him to tender

This letter is only suitable for use with WCD 98

尊敬的先生：

谨提及你方于［填入日期］的来信，信中表示有意对上述项目投标，现欣然附上下列文件：

1. 《雇主要求》两份。
2. 投标表格两份。
3. 填好地址的信封一个，以便寄回标书及与之相关的说明材料。

请注意以下事项：

a. 请与［填入姓名，电话］联系安排工地调查。
b. 投标将遵循《1996 年设计施工招投标程序规范》进行。

请将填好的投标表格装入附上的信封中封好，在［填入日期］［填入时间］之前寄到［填入地方］。

安全收到此函及附件后请告知我方，并请确认你方将按此说明提交标书。

你忠诚的

抄送：雇主

Dear Sir

I refer to your letter dated [*insert date*] in which you expressed a willingness to tender for the above project. I now have pleasure in enclosing the following:

1. Two copies of the Employer's Requirements.
2. Two copies of the form of tender.
3. An addressed envelope for the return of the tender and instructions relating thereto.

Please note the following:

a. The site may be inspected by arrangement with [*insert the person and telephone number*].
b. Tendering will be in accordance with the Code of Procedure for Selective Tendering for Design and Build 1996.

The completed form of tender is to be sealed in the endorsed envelope provided and must arrive at [*insert place*] not later than [*insert time*] on [*insert date*].

Please acknowledge safe receipt of this letter together with the enclosures noted and confirm that you will submit a tender in accordance with these instructions.

Yours faithfully

Copy: Employer

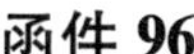

函件 96

致委托人和工料测量师，随函附上投标邀请书副本

Letter 96

To client and quantity surveyor, enclosing copy of invitation to tender

尊敬的先生：

我方已经于今日向以上项目投标人名单上的所有承包商寄去了投标文件。随函附上我方去函的副本，供你方参考。

你忠诚的

Dear Sir

I have today sent tender documents to all the contractors on the agreed list of tenderers for the above project. A copy of my covering letter is enclosed for your information.

Yours faithfully

函件 97

关于投标过程中的问题，致所有承包商

Letter 97

To all contractors, regarding questions during the tender period

尊敬的先生：

谨提及于［填入日期］寄去的以上项目的投标邀请书。

下列是收到的所有问题及今天做出的回答：

［简洁明了地列出问题及回答］

以上信息和澄清事项应被视为投标文件的一部分，并制约投标文件。

［视情况，加上：］

请注意，考虑到以上的修改，接收标书的日期已改为［填入新日期］。收取时间和地点不变。

你忠诚的

Dear Sir

I refer to the invitation to tender sent to you on [*insert date*] in respect of the above project.

The following is a list of all the questions which have been asked and the replies given at today's date:

[*List questions and answers clearly and concisely*]

The above information and clarification is to be taken as part of, and will override, the tender documentation as indicated.

[*Add, if appropriate*:]

Please note that the date for receipt of tenders has been changed to [*insert new date*] to take account of the above amendments. The time and place for receipt of tenders remains unchanged.

Yours faithfully

函件 98

关于开启标书，致委托人

Letter 98

To client, regarding opening of tenders

尊敬的先生：

谨提及我们近期的电话谈话。现确认主合同标书将于［填入日期］在我方办公地点收取。收取标书截止时间为［填入时间］。

标书将在你方到达后，于［填入时间］开启。我方已经告知工料测量师［填入姓名］，他将到场协助评定标书。

期待与你方会面。

你忠诚的

Dear Sir

I refer to our recent telephone conversation and confirm that main contract tenders should be received at this office on [*insert date*]. The deadline for receipt of tenders is [*insert time*].

Tenders will be opened at [*insert time*] after you arrive. I have spoken to [*insert name*], the quantity surveyor, who has arranged to be present to assist in assessing the tenders.

I look forward to seeing you.

Yours faithfully

函件 99

致提交附有条件的标书的承包商

Letter 99

To contractor who submits a qualified tender

尊敬的先生：

以上项目的标书已于［填入日期］开启。我方注意到你方提交的标书为有条件标书。

我方于［填入日期］寄给你方的投标邀请书中说明投标将遵循《1996 年单一阶段招投标程序规范》进行。此规范第 4.4.3 条规定，禁止有条件的标书。

如果你方仍然希望我方考虑此标书，请于［填入日期］前书面通知我方，你方取消限定条件，并且标书的内容不变。否则，你方标书将被拒绝。

你忠诚的

Dear Sir

Tenders for the above project were opened on [*insert date*] and it was noted that you had submitted a qualified tender.

The invitation to tender sent to you on the [*insert date*] stated that tendering was to be in accordance with the Code of Procedure for Single Stage Selective Tendering 1996 which prohibits qualified tenders by clause 4.4.3.

If you wish your tender to be considered, please inform me in writing by [*insert date*] that you withdraw the qualification without amendment to your tender. Failure to so notify me will result in your tender being rejected.

Yours faithfully

函件 100

致标书报价为第二或第三最低价的承包商

Letter 100

To contractors who submit the second and third lowest tenders

尊敬的先生：

以上项目的标书已经于［填入日期］开启，现告知你方标书为报价第二最低价/第三最低价［视情况取舍］。

尽管不是最有利，但如果我方进一步考虑你方的标书，我方将再次通知你方。当然，如果我方决定接受其他标书，我方也会立刻通知你方。

你忠诚的

抄送：雇主

工料测量师

Dear Sir

Tenders for the above project were opened on the [*insert date*] and I have to inform you that your tender was second/third [*delete as appropriate*] lowest.

Although not the most favourable, it may be that your tender will be given further consideration, in which case I will notify you again. I will, of course, notify you immediately if a decision is taken to accept another tender.

Yours faithfully

Copies: Employer

Quantity surveyor

函件 101

致标书报价不在前三名最低价之列的投标人

Letter 101

To contractors not among the three lowest tenderers

尊敬的先生：

以上项目的标书已经于［填入日期］开启，现通知你方，你方的标书报价不在前三名最低价之列。

虽然此次投标未中，但请你方确信，此次投标不会影响我方将来向你方发出其他问询。

我方将随后寄发所有标书报价列表。

你忠诚的

Dear Sir

Tenders for the above project were opened on the [*insert date*] and I have to inform you that your price was not among the lowest three submitted.

Although you were not successful in this instance, you can be sure that it will not adversely affect future enquiries from this office.

A full list of all the prices submitted will be sent to you in due course.

Yours faithfully

函件 102

当委托人已接受其他标书时，致标书报价为第二或第三低的承包商

Letter 102

To contractors who submit the second and third lowest tenders if another tender accepted

尊敬的先生：

谨提及我方于［填入日期］的去函。

我方的委托人已经决定接受另一份标书。感谢你方为实施你方的标书所做的准备，虽然此次投标未中，但请你方确信，此次投标不会影响我方将来向你方发出其他问询。

我方将随后向你方寄发所有标书报价列表。

你忠诚的

抄送：雇主
工料测量师

Dear Sir

I refer to my letter of the [*insert date*].

My client has decided to accept another tender. I thank you for being prepared to stand by your tender and, although you were not successful in this instance, you can be sure that it will not adversely affect future enquiries from this office.

A full list of all the prices submitted will be sent to you in due course.

Yours faithfully

Copies: Employer
Quantity surveyor

函件 103

致标书报价最低的承包商，但标书中出现错误，需按方案 1 进行处理

Letter 103

To contractor who submits the lowest tender, but with errors to be dealt with under Alternative 1

尊敬的先生：

我方工料测量师已经审查了你方关于以上项目的报价清单，发现如下错误：

［按页数和条款号列出错误］

按照《1996 年单一阶段招投标程序规范》第 6 条的方案 1，请你方向我方发出书面通告，确认或撤回你方的报价。

如果你方确认报价，我方将在报价清单上背书，标明填入的所有费率或价格（不包括初步条款、意外事件、最初造价和暂定总金额）将被视为是按修正后的价格总额少于/多于［视情况取舍］填入的报价总额的相同比例降低/提高后［视情况取舍］的价格。

你忠诚的

抄送：雇主

工料测量师

Dear Sir

The quantity surveyor has now completed his examination of your priced bills in connection with the above project and he has detected the following errors:

[*List errors by page and item number*]

In accordance with Alternative 1 of clause 6 of the Code of Procedure for Single Stage Selective Tendering 1996, you may now send me written notice confirming or withdrawing your offer.

If you confirm your offer, an endorsement will be added to the priced bills indicating that all rates or prices (excluding preliminary items, contingencies, prime cost and provisional sums) inserted therein are to be considered as reduced/increased [*delete as appropriate*] in the same proportion as the corrected total of priced items falls short of/exceeds [*delete as appropriate*] such items.

Yours faithfully

Copies: Employer
Quantity surveyor

函件 104

致标书报价最低的承包商，但标书中出现错误，需按方案 2 进行处理

Letter 104

To contractor who submits lowest tender, but with errors to be dealt with under Alternative 2

尊敬的先生：

我方工料测量师已经审查了你方关于以上项目的报价清单，发现如下错误：

［按页数和条款号列出错误］

按照《1996 年单一阶段招投标程序规范》第 6 条的方案 2，请你方确认你方报价或修改其中错误。

如果你方确认报价，我方将在报价清单上背书，标明填入的所有费率或价格（不包括初步条款、意外事件、最初造价和暂定总金额）将被视为是按修改后的价格总额少于/多于［视情况取舍］填入的报价总额的比例降低/提高后［视情况取舍］的价格。如果你方修改标书报价，并且此报价不再是最低报价，那么我方将考虑最低价标书。

你忠诚的

抄送：雇主

工料测量师

Dear Sir

The quantity surveyor has now completed his examination of your priced bills in connection with the above project and he has detected the following errors:

[*List errors by page and item number*]

In accordance with Alternative 2 of clause 6 of the Code of Procedure for Single Stage Selective Tendering 1996, you may now confirm your offer or amend it to correct the errors.

If you confirm your offer, an endorsement will be added to the priced bills indicating that all rates or prices (excluding preliminary items, contingencies, prime cost and provisional sums) inserted therein are to be considered as reduced/increased [*delete as appropriate*] in the same proportion as the corrected total of priced items falls short of/exceeds [*delete as appropriate*] such items. If you amend your offer and it is no longer lowest, the lowest tender will be considered.

Yours faithfully

Copies: Employer
Quantity surveyor

函件 105a

致承包商，接受标书须以正式文件为准

专递/挂号邮件

Letter 105a

To contractor, accepting tender subject to formal documents

Special/recorded delivery

尊敬的先生：

按我方委托人［填入姓名］的指示，现通知你方，依照编号为［填入编号］的工程图纸和工程量清单［或工程详细说明］，接受你方［填入日期］发出的关于以上项目的总金额为［填入金额］的标书，但需要以签订正式文件为准。

在今天的电话商谈中你方已经同意，进驻工地的日期为［填入日期］，相应的竣工日期为［填入日期］。

我方正在准备正式文件，并将于几天内寄发你方作为契约签署/执行［视情况取舍］。

你忠诚的

抄送：雇主

工料测量师

Dear Sir

My client [*insert name*] has instructed me to inform you that he accepts your tender dated [*insert date*] in the sum of [*insert amount in words*] for the above work in accordance with drawings numbers [*insert numbers*] and the bills of quantities [*or specification*] subject to the execution of formal documents.

As agreed by telephone today, the date for possession will be [*insert date*], and consequently the date for completion will be [*insert date*].

Formal documents are being prepared and they will be forwarded to you for signing/executing as a deed [*delete as appropriate*] within the next few days.

Yours faithfully

Copies: Employer
Quantity surveyor

函件 105b

致承包商，接受标书，立即建立合同

专递/挂号邮件

Letter 105b

To contractor, accepting tender and forming contract immediately

Special/recorded delivery

尊敬的先生：

按我方委托人［填入姓名］的指示，现通知你方，依照编号为［填入编号］的工程图纸和工程量清单［或工程详细说明］以及清单开始部分中注明的合同条款，接受你方［填入日期］发出的关于以上项目的总金额为［填入金额］的标书。

在今天的电话商谈中你方已经同意，进驻工地的日期为［填入日期］，相应的竣工日期为［填入日期］。

你忠诚的

Dear Sir

My client [*insert name*] has instructed me to inform you that he accepts your tender dated [*insert date*] in the sum of [*insert amount in words*] for the above work in accordance with the drawings numbers [*insert numbers*], the bills of quantities [*or specification*] and the terms of the contract noted in the preliminaries section of the bills of quantities [*or specification*].

As agreed by telephone today, the date for possession will be [*insert date*], and consequently the date for completion will be [*insert date*].

Yours faithfully

函件 106

关于中标人，致顾问

Letter 106

To consultants, regarding successful tenderer

尊敬的先生：

标书已经于［填入日期］开启，中标人为［填入姓名、地址］。

工地施工预期于［填入日期］开始，你方应谨记于此日期前完成你方所有的准备工作。

我方于［填入日期］安排了一次会议，届时承包商将会到场。请你方确认届时会出席会议。

你忠诚的

Dear Sir

Tenders were opened on the [*insert date*] and the successful tenderer was [*insert name*] of [*insert address*].

Work on site is expected to commence on [*insert date*] and you should finalise all your preparations with this date in mind.

I have arranged a meeting, at which the contractor will be present, on the [*insert date*]. Please confirm that you will attend.

Yours faithfully

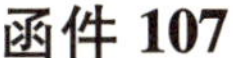

函件 107

致未中标者

Letter 107

To unsuccessful tenderers

尊敬的先生：

谨提及我方于［填入日期］的信函，并确认你方对以上项目的标书未中标。

收到的标书如下：

［按渐增顺序排列标书金额，只要不存在使投标人和金额相对应的可能，就可以另外按字母顺序列出投标人名称］。

你忠诚的

Dear Sir

I refer to my letter of the [*insert date*] and I confirm that your tender for the above work was not successful.

Tenders received were as follows:

[*List amounts in ascending order. Names of tenderers may also be listed, separately, in alphabetical order unless to do so would make it possible to link tenderers with amounts.*]

Yours faithfully

第九章　项目规划

谨记要自己准备合同文件。不要让工料测量师或委托人的律师来做此项工作。尽管有时候不可避免地要用到合同意向书，但使用它可能会有不少风险。合同意向书要措辞恰当，否则有可能造成一份有效的合同被终止。写完合同意向书后，应请专家就信函草稿给予建议。

核查并确定所有保险都已经由相应的合同各方进行了投保。建立核查清单，尤其是那些合同中没有规定的，需要按照雇主意愿来投保的保险。要确定承包商获得任何业绩、预付款和工地外材料保证金的途径都是遵循合同条款的。一个不错的办法就是在合同中预先加入一个条件：只有承包商付了保证金之后，合同才可以执行。

在这一阶段，要完成所有必要的准备工作，以便工地开工。你应该借此机会向承包商确认工程监督员的职责。否则以后可能会引起麻烦。这一阶段要用到一些标准信函来处理应该在K阶段之前确定的常规事项。

函件 108

致承包商，随函附上合同文件

专递/挂号邮件

Letter 108

To contractor, enclosing the contract documents

Special/recorded delivery

尊敬的先生：

谨提及我方于［填入日期］发出的信函，信函通知你方，我方接受你方对以上项目的投标［视情况，加上"以执行正式文件为准"］。现欣然附上两份合同文件，如下：

1. ［填入某种合同格式名称，包括所有官方的合同修正案编号和修改年以及补充条款。］

2. 工程概算表/工程详细说明/《雇主要求》/《承包商建议书》/《合同总金额分析》［视情况取舍］。

3. ［任何类似文件，如减量清单。］

4. 编号为［填入编号］的工程图纸。

请仔细审查合同文件，然后：

a. ［如果作为契约由一个法人实体执行］必须由两位董事，或一位董事和一位公司秘书在打印合同表的相应证明条款中签字。［如果作为契约由个人执行］须在有证人在场的情况下签署打印合同表中相应的证明条款。［如果作为盖印契约］须在有证人在场的情况下签署打印合同表中相应的证明条款，并在标明的位置盖上你方的印章。［如果作为非盖印契约］须在有证人在场的情况下签署打印合同表中相应的证明条款。

b. 在指定处草签打印合同表。

c. 有证人在场的情况下，在工程概算表［或工程详细说明，或《雇主要求》、《承包商建议书》和《合同总金额分析》］以及每份工程图纸上标明的位置签字。

请将所有文件寄还给我方，在雇主签字之后，我方将寄回一套给你方。

你忠诚的

Dear Sir

I refer to my letter of the [*insert date*] notifying acceptance of your tender for the above work [*if appropriate add*: '*subject to the execution of formal documents*']. I now have pleasure in enclosing two copies of the contract documents as follows:

1. [*Insert name of particular form of building contract together with all the official amendment numbers and years and any supplements.*]
2. Priced bills of quantities/specification/Employer's Requirements/Contractor's Proposals/Contract Sum Analysis [*delete as appropriate*].
3. [*Any similar document such as a reduction bill.*]
4. Drawings numbers [*insert numbers*].

Please examine the documents carefully, then:

a. [*If executed as a deed by a corporate body*:] Two directors, or one director and the company secretary, must sign the appropriate attestation clause in the printed form of contract. [*If executed as a deed by an individual*:] Sign the appropriate attestation clause in the printed form of contract and have the signature witnessed. [*If a deed under seal*:] Sign and have witnessed the appropriate attestation clause in the printed form of contract and fix your seal in the place indicated. [*If under hand*:] Sign and have witnessed the appropriate attestation clause in the printed form of contract.
b. Initial the printed forms of contract where indicated.
c. Sign and have witnessed the priced bills [*or specification, or Employer's Requirements and Contractor's Proposals and the Contract Sum Analysis*] and each of the drawings as indicated.

Please return all the documents to me and, after completion by the employer, one set will be returned to you.

Yours faithfully

函件 109

致委托人，随函附上合同文件

专递/挂号邮件

Letter 109

To client, enclosing the contract documents

Special/recorded delivery

尊敬的先生：

1. ［填入某种合同格式名称，包括所有官方的合同修正案编号和修改年以及补充条款］。

2. 工程概算表/工程详细说明/《雇主要求》/《承包商建议书》/《合同总金额分析》［视情况取舍］。

3. ［任何类似文件，如减量清单］。

4. 编号为［填入编号］的工程图纸。

请仔细审查合同文件，然后：

a. ［如果作为契约由一个法人实体执行］必须由两位董事，或一位董事和一位公司秘书在打印合同表的相应证明条款中签字。［如果作为契约由个人执行］须在有证人在场的情况下签署打印合同表中相应的证明条款。［如果作为盖印契约］须在有证人在场的情况下签署打印合同表中相应的证明条款，并在标明的位置盖上你方的印章。［如果作为非盖印契约］须在有证人在场的情况下签署打印合同表中相应的证明条款。

b. 在指定处草签打印合同表。

c. 有证人在场的情况下，在工程概算表［或工程详细说明，或《雇主要求》、《承包商建议书》和《合同总金额分析》］以及每份工程图纸上标明的位置签字。

请将所有合同文件寄还给我方，以便标注日期。

你忠诚的

Dear Sir

I have pleasure in enclosing two copies of the contract documents as follows:

1. [*Insert name of particular form of building contract together with all the official amendment numbers and years and any supplements.*]
2. Priced bills of quantities/specification/Employer's Requirements/Contractor's Proposals/Contract Sum Analysis [*delete as appropriate*].
3. [*Any similar document such as a reduction bill.*]
4. Drawings numbers [*insert numbers*].

Please examine the documents carefully, then:

a. [*If executed as a deed by a corporate body*:] Two directors, or one director and the company secretary must sign the appropriate attestation clause in the printed form of contract. [*If executed as a deed by an individual*:] Sign the appropriate attestation clause in the printed form of contract and have the signature witnessed. [*If a deed under seal*:] Sign and have witnessed the appropriate attestation clause in the printed form of contract and fix your seal in the place indicated. [*If under hand*:] Sign and have witnessed the appropriate attestation clause in the printed form of contract.
b. Initial the printed form of contract where indicated.
c. Sign and have witnessed the priced bills [*or specification, or Employer's Requirements and Contractor's Proposals and Contract Sum Analysis*] and each of the drawings as indicated.

Please return all the documents to me so that I can date them.

Yours faithfully

函件 110

致承包商，返还一份合同文件副本

Letter 110

To contractor, returning one copy of the contract documents

尊敬的先生：

我方现欣然附上雇主按要求完成的所有合同文件各一份，供你方存档。

你忠诚的

Dear Sir

I have pleasure in enclosing one copy of the contract documents, duly completed by the employer, for your retention.

Yours faithfully

函件 111

关于致承包商的合同意向书，致委托人

Letter 111

To client, regarding letter of intent to contractor

尊敬的先生：

按你方［填入日期］的指示，现随函附上一份我方建议寄给中标人的合同意向书草稿。

为避免延误，希望你方能尽快同意。如果你方有任何指示，请于［填入日期］之前，电话联系我方。对此我方将不胜感激。

你忠诚的

Dear Sir

Following your instructions of the [*insert date*] I enclose a draft letter of intent which I propose to send to the successful tenderer.

I should be pleased to have your agreement as soon as possible to avoid delay. If you have any observations, I should be grateful if you would telephone me before [*insert date*].

Yours faithfully

函件 112

致承包商：合同意向书

Letter 112

To contractor：letter of intent

尊敬的先生：

受我方委托人［填入姓名］的指示，现通知你方，你方于［填入日期］对于［填入工程性质］的总金额为［填入金额］的标书可以接受，我方准备起草主合同文件，以便双方签署。合同须以我方委托人［填入适当的附加条件］为准。

我方的委托人无意将此函作为有约束力的合同的证明。不过，我方的委托人准备指示你方进行［填入有限的工程内容细节］。无论因为何种原因，如果对此工程没有与你方签订合同，我方委托人的责任就只限于通过主承包商对［填入有限的工程内容］付款。

如果没有书面订单，你方不需要进行标书中的其他工程。在任何情况下，我方的委托人和我方都不承担任何其他责任。

你忠诚的

Dear Sir

My client [*insert name*], has instructed me to inform you that your tender of the [*insert date*] in the sum of [*insert amount*] for [*insert the nature of the work*] is acceptable and that I intend to prepare the main contract documents for signature subject to my client [*insert the appropriate provisos*].

It is not my client's intention that this letter should be evidence of a binding contract. However, my client is prepared to instruct you to [*insert details of the limited nature of the work required*]. If, for any reason whatsoever, no contract is entered into with you for this work, my client's commitment will be strictly limited to payment for [*insert the limited nature of the work required*].

No other work included in your tender must be carried out without a further written order. No further obligation is placed upon my client and no obligation whatsoever, under any circumstances, is placed upon me.

Yours faithfully

函件 113a

关于保险事宜，致承包商

此信函不适用于 MW98 合同和 GC/Works/1（1998）合同

Letter 113a

To contractor, regarding insurance

This letter is not suitable for use with MW 98 or GC/Works/1（1998）

尊敬的先生：

我方代表我方的雇主，要求你方提交合同第 21.1.2 条［使用 IFC98 合同时，替换为“第 6.2.2 条”］规定的保险单和保险费收据以便雇主检查。这些文件须于［填入日期］之前寄到我处。

［视情况，加上：］

要求你方提交你方准备按合同第 22A 条［使用 IFC98 合同时，替换为“第 6.3 条”］投保的保险人名称，以取得雇主的批准。此信息必须于［填入日期］之前送达我处。

［视情况，加上：］

须遵循第 21.2.1 条进行投保［使用 IFC84 合同时，替换为“第 6.2.4 条”］。现特此指示你方以共同名义投保，保额为附件中说明的赔偿金额。在投保之前，请提交你方计划投保的保险人名称，以取得雇主批准。

你忠诚的

抄送：雇主

Dear Sir

I write on behalf of the employer to request you to submit to me insurance policies and premium receipts in accordance with clause 21. 1. 2 [*substitute* '*6. 2. 2*' *when using IFC 98*] of the contract for inspection by the employer. These documents must be in my hands by no later than [*insert date*].

[*Add*, *if appropriate*:]

You are further requested to submit to the employer for his approval the name of the insurers with whom you propose to take out insurance under clause 22A [*substitute* '*6. 3A*' *when using IFC 98*]. This information must be in my hands no later than [*insert date*].

[*Add*, *if appropriate*:]

Insurance will be required in accordance with clause 21. 2. 1 [*substitute* '*6. 2. 4*' *when using IFC 84*] and we hereby instruct you to take out and maintain a joint names policy for the amount of indemnity specified in the appendix. Before so doing, please submit to me, for approval by the employer, the name of the insurers with whom you propose to take out such insurance.

Yours faithfully

Copy: Employer

函件 113b

关于保险事宜，致承包商

此信函仅适用于 MW98 合同

Letter 113b

To contractor, regarding insurance

This letter is only suitable for use with MW 98

尊敬的先生：

我方代表我方的雇主，要求你方提交保险单和保险费收据，以证明合同第 6.1 条和 6.2 条规定的保险已经投保并生效。所有的文件，包括分包商的保险单和保险费收据在内，必须于［填入日期］之前寄到我处。

［视情况，加上：］

同时，要求你方提交保险单和保险费收据，以证明合同第 6.3A 条规定的保险已经投保并生效。注意，这些保险必须以雇主和你方的共同名义投保，保额为全部重建价值加上第 6.3A 条中规定的额外的［填入百分比］以赔付专业费用。这些文件必须于［填入日期］之前寄到我处。

你忠诚的

抄送：雇主

Dear Sir

I write on behalf of the employer to request you to submit to me insurance policies and premium receipts to show that the insurances referred to in clauses 6. 1 and 6. 2 have been taken out and are in force. These documents, which should include policies and premium receipts in respect of all subcontractors, must be in my hands no later than [*insert date*].

[*Add, if appropriate*:]

You are further requested to submit insurance policies and premium receipts to show that the insurances referred to in clause 6. 3A have been taken out and are in force. Note that such insurances must be taken out in the joint names of the employer and yourself and must be for the full reinstatement value plus [*insert percentage*] as stated in clause 6. 3A to cover professional fees. These documents must be in my hands no later than [*insert date*].

Yours faithfully

Copy: Employer

函件 113c

关于保险事宜，致承包商

此信函仅适用于 GC/Works/1（1998）合同

Letter 113c

To contractor, regarding insurance

This letter is only suitable for use with GC/Works/1 (1998)

尊敬的先生：

我方代表我方的雇主要求你方提交：

[加上：]

所有合同第8（2）条和方案A第8（3）条规定的关于雇主责任的保险，合同条款规定的关于你方承担的工程及物品丢失或损坏的保险，以及工程中或与工程相关的任何人的人身伤害保险和财产丢失或损坏保险的保险单副本。

[或：]

所有合同第8（2）条和方案B第8（3）条规定的关于雇主责任的保险，以及专用条款后附加的基本保险要求总结中所规定的保险项目的保险单副本。

你忠诚的

Dear Sir

I write on behalf of the employer to request you to submit to me
[*Add either*:]

copies of insurances in respect of employer's liability, loss or damage to the Works and things for which you are responsible under the terms of the contract and insurance against personal injury to any persons and loss or damage to property arising from or in connection with the Works, all in accordance with clause 8 (2) and alternative A clause 8 (3) of the contract.

[*Or*:]

copies of insurance in respect of employer's liability and insurances in accordance with the summary of essential insurance requirements attached to the abstract of particulars, all as clause 8 (2) and alternative B clause 8 (3) of the contract.

Yours faithfully

函件 114a

关于保险事宜，致委托人

此信函不适用于 MW98 或 GC/Works/1（1998）合同

Letter 114a

To client, regarding insurance

This letter is not suitable for use with MW 98 or GC/Works/1 (1998)

尊敬的先生：

随函附上寄给承包商的关于保险的信函。信函内容不释自明。我方不具有保险方面的专业知识，所以承包商提交其所需的信息时，你方应该立即转给你方的保险经纪人以征询意见。如果你方能请经纪人将他的意见寄给我方一份，我方将转给承包商。

［视情况，加上：］

按合同第 22B/22C 条［视情况取舍，使用 IFC98 合同时，替换为“第 6.3B 条”或“第 6.3C 条”］要求你方和承包商以共同名义投保所有保险。按附件中的说明，投保金额为工程的完全重建价值加上额外的［填入百分比］以赔付专业费用，［视情况加上］同时双方以共同名义对现有结构和内部财物存在的特定危险进行投保。在特定情况下，有可能扩展现有保险责任范围。随函附上合同中相关部分摘录，请你方立即与你方的保险经纪人讨论此事，以便保险从［填入日期］开始生效。如果你方安排好了保险事宜，请告知我方，以便我方及时通知承包商。

你忠诚的

Dear Sir

I enclose a letter which I sent to the contractor regarding insurances. The contents are self-explanatory. I have no expertise in insurance matters. Therefore, when the contractor submits the requested information, you should pass it immediately to your insurance broker for his comments. If you will ask your broker to send a copy of his comments to me, I will pass them to the contractor.

[*Add, if appropriate*:]

The contract requires you to take out all risks insurance cover in the joint names of the contractor and yourself under clause 22B/22C [*delete as appropriate and substitute '6.3B' or '6.3C' when using IFC 98*] for the full reinstatement value of the Works plus [*insert percentage*] as stated in the appendix to cover professional fees [*add, if appropriate*] together with joint names insurance against specified perils for existing structures and contents. In certain circumstances, it may be possible to extend your existing insurance cover. I enclose the relevant extracts from the contract and you should speak to your own insurance broker without delay so that cover is effective from [*insert date*]. Please let me know when you have arranged cover so that I can notify the contractor.

Yours faithfully

函件 114b

关于保险事宜，致委托人

此信函只适用于 MW98 合同

Letter 114b

To client，regarding insurance

This letter is only suitable for use with MW 98

尊敬的先生：

随函附上寄给承包商的关于保险事宜的信函。信函内容不释自明。我方不具有保险方面的专业知识，所以承包商提交其所需的信息时，你方应该立即转给你方的保险经纪人以征询意见。如果你方能请经纪人将他的意见寄给我方一份，我方将转给承包商。

［视情况，加上：］

按合同第 6.3B 条，要求你方和承包商以共同名义对现有结构和内部财务及工程的损失或毁坏投保。尽管越来越困难，在特定情况下，仍然有可能扩展现有的保险责任范围。随函附上合同中相关部分摘录，请你方立即与你方的保险经纪人讨论此事，以便保险从［填入日期］开始生效。如果你安排好了保险事宜，请告知我方，以便我方及时通知承包商。

你忠诚的

Dear Sir

I enclose a letter which I have sent to the contractor regarding insurances. The contents are self-explanatory. I have no expertise in insurance matters. Therefore, when the contractor submits the requested information, you should pass it immediately to your own insurance broker for his comments. If you will ask your broker to send a copy of his comments to me, I will pass them to the contractor.

[*Add, if appropriate:*]

The contract requires you to take out insurance cover in the joint names of the contractor and yourself under clause 6.3B against loss or damage to the existing structures and contents, and to the Works. In certain circumstances, it may be possible to extend your existing insurance cover, although this is becoming more difficult. I enclose the relevant extracts from the contract and you should speak to your broker without delay so that cover is effective from [*insert date*]. Please let me know when you have arranged cover so that I can notify the contractor.

Yours faithfully

函件 114c

关于保险事宜，致委托人

此信函只适用于 GC/Works/1 （1998） 合同

Letter 114c

To client, regarding insurance

This letter is only suitable for use with GC/Works/1 (1998)

尊敬的先生：

随函附上寄给承包商的关于保险事宜的信函。信函内容不释自明。我方不具有保险方面的专业知识，所以承包商提交其所需的信息时，你方应该立即转给你方的保险经纪人以征询意见。如果你方能请经纪人致函给我方，我方将把他的意见转给承包商。

你忠诚的

Dear Sir

I enclose a letter which I have sent to the contractor regarding insurances. The contents are self-explanatory. I have no expertise in insurance matters. Therefore, when the contractor submits the requested information, you should pass it to your own insurance broker for his comments. If you will ask your broker to write directly to me, I will pass the comments to the contractor.

Yours faithfully

函件 115

关于保险单事宜，致承包商

此信函不适用于 MW98 或 GC/Works/1（1998）合同

Letter 115

To contractor, regarding insurance policies

This letter is not suitable for use with MW 98 or GC/Works/1 (1998)

尊敬的先生：

我方现寄还关于合同第 21.1.2 条［使用 IFC98 合同时，替换为“第 6.6.2 条”］中所涉及的保险单［填入编号］及保险费收据［填入编号］。雇主已检查完毕，［视情况，加上：］以下为雇主意见：［填入从雇主保险经纪人处收到的意见］

［视情况，加上：］

你方按第 22A 条［使用 IFC98 合同时，替换为“第 6.3A 条”］规定向［填入被建议的投保人名称］投保的建议书已经雇主批准。注意这些保险必须以雇主和你方的共同名义投保，投保金额必须为完全重建价值加上额外的［填入百分比］，以赔付专业费用。保险单据和已付保险费收据必须于［填入日期］之前寄于雇主。

［视情况，加上：］

你方按第 21.2.1 条［使用 IFC98 合同时，替换为“第 6.2.4 条”］规定向［填入被提议的投保人名称］投保的建议书已经雇主批准。请以雇主和你方的共同名义投保，投保额为附件中说明的赔偿金额。保险单据和已付保险费收据必须于［填入日期］之前寄于雇主。

你忠诚的

Dear Sir

I return herewith the insurance policies numbers [*insert numbers*] and premium receipts numbers [*insert numbers*] in respect of clause 21. 1. 2 [*substitute* '*6. 2. 2*' *when using IFC 98*] of the contract. The employer has completed his inspection [*add*, *if appropriate*:] and he has the following comments to make: [*insert the comments received from the employer's brokers*].

[*Add*, *if appropriate*:]

Your proposal to take out insurance with [*insert name of proposed insurers*] in accordance with the requirements of clause 22A [*substitute* '*6. 3A*' *when using IFC 98*] is approved by the employer. Note that such insurances must be taken out in the joint names of the employer and yourself and must be for the full reinstatement value plus [*insert percentage*] as stated in the appendix to cover professional fees. The policy or policies and receipts for premiums paid must be deposited with the employer no later than [*insert date*].

[*Add*, *if appropriate*:]

Your proposal to take out insurance with [*insert name of proposed insurers*] in accordance with the requirements of clause 21. 2. 1 [*substitute* '*6. 2. 4*' *when using IFC 98*] is approved by the employer. Please proceed to take out insurance in the joint names of the employer and yourself for the amount of indemnity specified in the appendix. The policy or policies and receipts for premiums paid must be deposited with the employer no later than [*insert date*].

Yours faithfully

函件 116

关于预定违约赔偿保险事宜，致承包商

此信函不适用于 MW98 或 GC/Works/1（1998）合同

Letter 116

To contractor, regarding liquidated damages insurance

This letter is not suitable for use with MW 98 or GC/Works/1 (1998)

尊敬的先生：

尽管附件中说明有可能要求第 22D 条［使用 IFC98 合同时，替换为“第 6.3D 条”］保险，现我方确认不需要此项保险。

［或：］

第 22D 条［使用 IFC98 合同时替换为“第 6.3D 条”］中涉及到的保险需按附件中提到的期限投保，欣然希望你方能拿到其报价。此保险的价格须经双方都同意。如果你方合理地需要任何信息以取得报价，请立即通知我方。

你忠诚的

Dear Sir

[*Either*:]

Although it is stated in the appendix that clause 22D [*substitute '6.3D' when using IFC 98*] insurance may be required, I confirm that no such insurance is required.

[*Or*:]

I should be pleased if you would obtain a quotation for insurance to which clause 22D [*substitute '6.3D' when using IFC 98*] refers for the time period stated in the appendix. The insurance must be on an agreed value basis. Please let me know without delay if you reasonably require any further information to obtain such a quotation.

Yours faithfully

函件 117

收到预定违约保险报价后，致承包商

此信函不适用于 MW98 或 GC/Works/1（1998）合同

Letter 117

To contractor, after receiving liquidated damages insurance quotation

This letter is not suitable for use with MW 98 or GC/Works/1 (1998)

尊敬的先生：

你方转寄来的关于第 22D 条［使用 IFC98 合同时替换为“第 6.3D 条”］中涉及的保险的报价单收悉，深表谢意。

［然后：］

我方已收到雇主指示，不希望你方接受此报价。

［或：］

我方已收到雇主指示，希望你方接受此报价。请你方按照相关的投保措施，着手办理保险，并连同保险费收据和任何相关背书（或其中的背书和相应的保险费收据）寄给我方，以便雇主保存。我方对此将不胜感激。

你忠诚的

Dear Sir

Thank you for forwarding the quotation you have received in connection with the insurance to which clause 22D [*substitute* '*6.3D*' *when using IFC 98*] refers.

[*Then*, *either*:]

I have received instructions from the employer that he does not wish you to accept the quotation in this instance.

[*Or*:]

I have received instructions from the employer that he wishes you to accept the quotation and I should be pleased if you would forthwith take out and maintain the relevant policy and send it to me for deposit with the employer together with the premium receipt therefor and any relevant endorsement or endorsements thereof and the premium receipts therefor.

Yours faithfully

函件 118

关于履约保证书，致承包商

Letter 118

To contractor, regarding performance bond

尊敬的先生：

合同清单［或工程详细说明］第［填入页码］第［填入参考条目］项要求你方提供总额为［填入金额］的履约保证书。

请你方立即准备保证书，完成后请寄给我方原件，而不是副本。此保证书将存放于雇主处，直到工程实际竣工/竣工/缺陷整改竣工［视情况取舍］。

你忠诚的

Dear Sir

Page [*insert number*] item [*insert item reference*] of the contract bills [*or specification*] requires you to provide a performance bond in the sum of [*insert amount*].

Please arrange the bond immediately and, on completion, let me have the original document, not a copy, which will be lodged with the employer until after practical completion/completion/completion of making good defects [*delete as appropriate*] of the Works.

Yours faithfully

函件 119

关于预付款保证书，致承包商

此信函不适用于 MW98 或 GC/Works/1（1998）合同

通过传真或邮寄

Letter 119

To contractor, regarding advance payment bond

This letter is not suitable for use with MW 98 or GC/Works/1（1998）

By fax and post

尊敬的先生：

附录中说明使用第 30.1.1.6 条［使用 IFC98 合同时，替换为“第 4.2（b）条”，使用 WCD98 合同时，替换为“第 30.1.1.3 条”］。同时附录中也明确说明需要预付款保证书。因此，在我方预付款之前，你方必须提供此保证书。

鉴于此，欣然希望你方能立即向雇主提交你方建议的保证人的姓名，以期得到雇主批准。

你忠诚的

抄送：雇主

Dear Sir

The appendix states that clause 30.1.1.6 [*substitute* ‘*4.2（b）*’ *when using IFC 98 or* ‘*30.1.1.3*’ *when using WCD 98*] applies. It also states that an advance payment bond is required. You must provide such bond before any advance payment can be made.

With that in mind, I should be glad if, by return, you will submit the name of the proposed surety to the employer for approval.

Yours faithfully

Copy: Employer

函件 120

关于工地外材料保证书，致承包商

此信函仅适用于 JCT98 或 IFC98 合同

通过传真或邮寄

Letter 120

To contractor, regarding off-site materials bond

This letter is only suitable for use with JCT 98 or IFC 98

By fax and post

尊敬的先生：

雇主已经向你方提供了一份物品清单。如果特定条件令人满意，这些物品在存放于工地之外时便可得到付款。清单已随合同清单附上。

证明书的条件为，对于合同附录中附加的雇主批准的保证人所不能证明的材料，你方必须提供一份保证书。

请于［填入日期］之前提交你方建议的保证人姓名。如果保证人得到了批准，需要在［填入日期］之前提交保证书。

你忠诚的

抄送：雇主

Dear Sir

The employer has supplied you with a list of items for which payment may be made while they are stored off-site if certain conditions are satisfied. The list is also attached to the contract bills.

It is a condition of certification that you must supply a bond in respect of items which are not uniquely identified from a surety approved by the employer on terms annexed to the appendix to the contract.

Please submit the name of your proposed surety by the [*insert date*]. If the surety is approved, the bond will be required by the [*insert date*].

Yours faithfully

Copy: Employer

函件 121

执行施工合同之初，致委托人

Letter 121

To client, at the beginning of the building contract

尊敬的先生：

谨借此机会，在承包商进驻工地开工之前，致函于你方。有几点值得强调，因为这几点会对工程合理而有效率地进行有很重要的影响，而且，这关系到你方作为合同雇主方的利益。

1. 在执行合同之前，我方的身份是你方的代理人，我方必须遵循我方的雇佣条款的相关规定。执行合同以后，尽管我方仍然是你方的代理人，但我方又多了一个额外的职责：在合同各方之间公平地管理合同。这意味着我方必须严格按合同条款做出决定。

2. 按合同要求，我方必须在合适的时间，依据我方自身的专业判断发出所有的证明书。其中一种重要的证明书就是规定了承包商应得的款项金额的财务证明。这些证明书在工程进行中要［填入频率］发一次，之后还要不时地发出。从每次证明书发出到向承包商付款之间，你方最多有［填入天数］天的时间期限。在规定的时间内按证明书上写明的金额付给承包商款项是极其重要的事。如果你方没能按规定时间付款，承包商则有权提前 7 天通知你方其暂停履行所有责任或终止雇佣关系。并且承包商有权按英国银行基本利率取得未付总金额的 5%。这样就会在资金和时间方面对合同造成毁灭性的后果。

3.《1996 年住房补贴、建造和改造法案》要求每一份建筑合同都应该包括某些特定的条款。这样做的一个重要结果就是对于我方的每一份财务证明，你方必须给承包商两份通知。第一份必须在我方发出财务证明［使用 WCD98 合同时，替换为“承包商申请”］后 5 天内以书面形式发出，说明你方计划付款的金额。如果你方希望拒付或减少付款金额，则要发出第二份通知，第二份通知需要在最终付款日期 5 天［使用 GC/Works/1（1998）合同时，替换为“7 天”］之前以书面形式发出。我方必须强调的是，发出通知是你方而不是我方的责任，但我方可以寄给你方合适的信函草稿，供你方使用。

（转下页）

Dear Sir

I am taking the opportunity to write to you before the contractor takes possession of the site and commences work. There are a number of points worth emphasising, because they have an important effect on the proper and efficient progress of the work and they concern you as employer under the contract:

1. Until the contract is executed, I am required to act solely as your agent within the limits laid down by the terms of my appointment. Thereafter, although I continue to act for you as before, I owe you the additional duty to administer the contract fairly between the parties. This means that I must make any decisions under the contract strictly in accordance with the terms of the contract.
2. The contract obliges me to issue all certificates at the appropriate time and to use my professional judgement in so doing. An important type of certificate I have to issue is the financial certificate which stipulates the amount of money due to the contractor. These certificates are issued at [*insert frequency*] intervals during the period that work is in progress and occasionally thereafter. You have a maximum of [*insert number*] days from the date of each certificate to get your payment to the contractor. The importance of paying the contractor the full amount indicated on the certificate within the period allowed cannot be stressed too much. Failure to honour certificates within the time limit allows the contractor to suspend performance of all his obligations on 7 days' notice and to commence the procedure for determining his employment and entitles him to interest on the outstanding sum at 5% over Bank of England base rate. The result of this would be disastrous to the contract in terms of both time and money.
3. The Housing Grants, Construction and Regeneration Act 1996 requires that certain clauses are to be included in every construction contract. An important result of this is that there are two notices which you must give to the contractor in respect of each of my financial certificates. The first notice must be given in writing no later than 5 days after the date of issue of my certificate [*substitute 'the contractor's application' when using WCD 98*], stating the amount you propose to pay. The second notice is to be sent if you wish to withhold or deduct any money from the amount due to be paid. This notice must be given in writing no later than 5 [*substitute '7' when using GC/Works/1 (1998)*] days before the final date for payment. I must emphasise that it is your responsibility, not mine, to send these notices, but I will send you some suitable draft letters which you may use for the purpose.

[*continued*]

函件 121 续表

Letter 121 continued

4. 如果承包商与你方通过信函、电话联系或是亲自拜访，请立即通知我方。不建议你方在没有与我方商议的情况下，就合同事宜回答承包商的任何询问或做出任何决定。这样做会代价昂贵。如果有任何事情需要你方决定，我方会立即转给你方，并附上我方的意见。

5. 如果你方希望不时地去工地查看工程进度情况，请通知我方，以便我方能抽出时间陪同你方前去，以处理任何可能出现的问题。当然，我方会将日常基本的工程进度与财务状况定期向你方汇报。

你忠诚的

4. If the contractor communicates with you by letter, telephone or personal visit, please refer him to me and let me know immediately. It is not advisable for you to answer any of his queries or make decisions regarding the contract without consulting me. It could be costly. If there are any matters requiring your decision, I will refer them to you as they arise, with my observations.

5. If you wish to visit the site from time to time, to see the work in progress, please let me know so that I can make myself available to accompany you on each occasion to take care of any points which may arise. I will, of course, keep you informed on a regular basis regarding the progress of the work and the financial situation.

Yours faithfully

函件 122

关于委任事宜，致工程监督员

Letter 122

To clerk of works, on appointment

尊敬的先生：

我方的委托人［填入姓名］已经向我方确认你方被委任为以上合同的工程监督员。欣然希望你能于［填入日期、时间］到我方的办公室，以便向你方简要介绍一下项目情况，并转交工程图纸、工程量清单［或工程详细说明］、周报告表及日记录本。

预计承包商将于［填入日期］进驻工地。希望你方于［填入期望工程监督员在场的时间段］能在场。按照承包商的工程计划，所有的工地临时住所应该在第［填入天数］之前完工。欣然希望你方能检查其是否符合合同规定。

合同第 12 条［使用 IFC98 合同时，替换为“第 3.10 条”，使用 GC/Works/1（1998）合同时，替换为“第 4 条”或“工程详细说明”］详细描述了你方的职责，现附上供你方参考。尤其希望你方注意以下几点：

1. 希望你方检查所有工艺和材料，以确定其符合合同要求。你方应该向相应负责人［使用 GC/Works/1（1998）合同时，替换为“代理人”］指出所有缺陷，并给出你方的意见。如果任何缺陷在 24 小时内没有得到修整或发生了重大与基础性的缺陷，你方必须电话通知我方。不要向承包商发出任何书面指示。

2. 尽管通常工程监督员会在工地缺陷工程上作出标记，但按合同规定，你方无权这样做。任何情况下，无论材料有没有被使用到建筑结构上，你方都不能在材料上涂写。

3. 通常在实际竣工［使用 GC/Works/1（1998）合同时，替换为“竣工”］之前，我方不会向承包商发出缺陷列表。此列表通常被称为“绊脚石清单”，因此有可能会被误解而引起争议。此列表应该由负责人［使用 GC/Works/1（1998）合同时，替换为“代理人”］来完成。请你方对承包商的评论仅限于口头意见。

4. 建筑师是惟一有权对承包商发出书面指示的人。

5. 你方无权变动工程、材料或设计。请把所有要求转告给我方。

6. 完成周报告表，尤其注意［按要求填入］并于每周一寄给我方。

7. 日记录要尽量完整。

（转下页）

Dear Sir

My client, [*insert name*], has confirmed your appointment as clerk of works for the above contract. I should be pleased if you would call at this office on [*insert date*] at [*insert time*] to be briefed on the project and to collect your copies of drawings, bills of quantities [*or specification*], weekly report forms and daily diary.

It is anticipated that the contractor will take possession of the site on the [*insert date*]. You are expected to be present on site [*insert periods during which the clerk of works is expected to be present*]. According to the contractor's programme, all site accommodation will be completed by day [*insert day number*] and I should be glad if you would check that it is in accordance with the contract.

Your duties will be as described in the contract clause 12 [*substitute '3. 10' when using IFC 98, '4' when using GC/VVorks/1 (1998) or 'specification' as appropriate*] a copy of which is enclosed for your reference. In particular, I wish to draw your attention to the following:

1. You will be expected to inspect all workmanship and materials to ensure conformity with the contract requirements. Any defects must be drawn to the attention of the person-in-charge [*substitute 'agent' when using GC/Works/1 (1998)*], to whom you should address all comments. If any defects are left unremedied for 24 hours or if they are of a major or fundamental nature, you must inform me immediately by telephone. Do not issue any written directions to the contractor.
2. Although it is common practice for clerks of works to mark defective work on site, the contract gives you no such power. You must not in any way deface materials on site whether or not they are incorporated into the structure.
3. It is not my policy to issue lists of defects to the contractor before practical completion [*substitute 'completion' when using GC/Works/1 (1998)*]. Commonly called 'snagging lists', they may be misinterpreted and give rise to disputes. They should be compiled by the person-in-charge [*substibute 'agent' when using GC/Works/*1 (1998)]. Please confine your remarks to the contractor to oral comments.
4. The architect is the only person empowered to issue instructions to the contractor.
5. You are not empowered to vary work or materials or design. Refer all queries to me.
6. Complete the weekly report sheets, paying especial attention to [*insert as required*] and send them to me each Monday.
7. Complete the diary as fully as possible.

[*continued*]

函件 122 续表

Letter 122 continued

8. 记住如果有争议发生，你方的周报告表和日记录可能会作为证据。做记录时应记住这一点。

9. 尽管我方可能会让你方做种种与合同有关的工作，你方的主要职责还是作为雇主的工程检查员。按合同条款，你方对承包商无任何职责，但按一般法律，你方不能做出任何不谨慎的声明。

希望你方能与承包商建立良好的关系，合同的成功执行很大程度上就取决于这种关系。如果对任何事情有疑问，请立即联系我方。

你忠诚的

8. Remember that your weekly report sheets and diary may be called in evidence in the case of a dispute. Bear this in mind when making entries.

9. Remember that, although I may call upon you to carry out various tasks in relation to the contract, your primary duty is to the employer as inspector on the Works. You have no duty to the contractor under the provisions of the contract, but you have a duty under the general law not to make careless statements.

I hope you achieve the kind of relationship with the contractor on which successful completion of the contract depends so much. Do not hesitate to contact me if you are in doubt about anything.

Yours faithfully

函件 123

致承包商，告知工程监督员的职权范围

Letter 123

To contractor, noting the authority of the clerk of works

尊敬的先生：

我方对以上工程已经委任了一名工程监督员［填入姓名］，他对此类工作经验丰富。我方希望未来几个月里，你们能建立起良好的关系。

工程监督员将会整周/每周几天［视情况取舍］在工地。合同第 12 条［使用 IFC98 合同时，替换为“第 3.10 条”，使用 GC/Works/1（1998）合同时，替换为“第 4 条”或“工程详细说明”］中规定了工程监督员的职责。他是工程材料和工艺质量的检验员，但无权发出指示，强调此点可能对你方有所帮助。尽管我方希望他能应相关要求，就工程中出现的任何问题给予建议，但按照合同进行施工仍是你方的责任。请注意工程监督员不能代替你方的管理人员。尽管工程监督员在我方指挥下工作，他只对雇主负有职责。

我方已经告知工程监督员，他不可以在工程上做标记来标出缺陷材料，因为这可能会引起麻烦。他不会发出任何“绊脚石清单”。这样可以避免对缺陷工程范围的误解。如果工程监督员发现任何不符合合同要求的工艺或材料，他会当场指出，并记录在日记录中向我方汇报。我方相信对此我方不需要发出具体的指示。当然，如果工程监督员由于工作失误而没有发现缺陷，则不会影响你方按合同履行责任。

如果你方对此信函内容有任何疑问，请立即致函或电话联系我方，以便得到明确解答。

你忠诚的

抄送：工程监督员

Dear Sir

A clerk of works has been appointed for the above contract. His name is [*insert name*] and he is experienced in work of this type. I hope that you will build up a successful relationship over the coming months.

The clerk of works will be on site during the whole/part of the week only [*delete as appropriate*]. His duties are as laid down in the contract clause 12 [*substitute '3.10' when using IFC 98, '4' when using GC/Works/1 (1998) or 'specification' as appropriate*]. I hope it will be helpful if I emphasise that he is acting as an inspector of materials and workmanship; he is not empowered to issue instructions. Although I expect that he will be ready to give his opinion, if requested, on any points which may arise during construction, the responsibility for carrying out the Works in accordance with the contract remains yours. Please note especially that the clerk of works is in no way a substitute for your own supervisory staff. The clerk of works owes his duty to the employer although he is under my direction.

I have informed him that he must not make any marks on the Works to indicate defective materials, because I know this can be a source of annoyance. No snagging lists will be issued. This should remove any misunderstandings regarding the extent of defective work. If the clerk of works notices any workmanship or materials not in accordance with the contract, he will point it out on site, note it in his diary and report it to me. I trust that it will be unnecessary for me to issue specific instructions regarding such matters. The failure of the clerk of works to notice defective work does not, of course, affect your own obligations under the contract.

If you are in any doubt regarding the contents of this letter, please do not hesitate to write or telephone me for clarification.

Yours faithfully

Copy: Clerk of works

函件 124

关于扩展工程监督员的职权范围，致承包商

Letter 124

To contractor, regarding extension of authority of the clerk of works

尊敬的先生：

欣然希望你方注意到，工程监督员将作为我方的授权代表负责，并仅负责合同第［填入条款编号］条。

［简要描述按该条款的工程监督员的职权范围］

工程监督员的职权范围的扩展适用到再次通知为止。但请注意，他的其他职责不受影响。

你忠诚的

抄送：工程监督员
工程量估算师

Dear Sir

I should be pleased if you would note that the clerk of works is my authorised representative for the purposes of clause［*insert clause number*］of the contract and for that purpose alone.

［*Set out a brief description of the authority of the clerk of works under the clause*］

This extension of the authority of the clerk of works will apply until further notice, but note that his other duties remain unaffected.

Yours faithfully

Copies: Clerk of works
Quantity surveyor

函件 125

致承包商，指定授权代表

Letter 125

To contractor, naming authorised representatives

尊敬的先生：

借此函正式通知你方，建筑师关于该合同的全部内容的授权代表为：

［列出代表的姓名及公司职务］

此安排适用到再次通知为止。

你忠诚的

谨代表［填入合同上的公司名称］。

抄送：雇主
顾问
工程监督员

Dear Sir

This is to formally let you know that the architect's authorised representatives for all the purposes of the contract are:

[*List representatives, giving names and positions in the firm*]

This arrangement will apply until further notice.

Yours faithfully

for and on behalf of
[*insert name in the contract which should be the firm name*]

Copies: Employer
Consultants
Clerk of works

函件 126

关于发出指示，致承包商

Letter 126

To contractor, regarding the issue of instructions

尊敬的先生：

现及时提请你方注意，建筑师是该合同中惟一经授权可发出指示的人。

除非经建筑师书面确认［使用 WCD98 合同时替换为“雇主代理”或“雇主”］，所有其他的无论是由谁发出的指示，都是无效的。

这一限制适用于雇主对此项目雇佣的所有顾问。尽管如此，如果你方收到建筑师以外的任何人发出的指示，你方应该及时转给我方。你方既不能按其指示施工，也不能无视此指示。

你忠诚的

抄送：雇主
工料测量师
顾问
工程监督员

Dear Sir

It seems timely to draw your attention to the fact that the architect is the only person authorised to issue instructions under this contract.

All other instructions, from whatsoever source, are of no effect unless confirmed by the architect [*substitute* '*employer's agent*' *or* '*employer*' *when using WCD 98*] in writing.

The restriction applies to all consultants engaged by the employer on this project. If, despite this letter, you receive an instruction from some person other than the architect, it should be referred to me immediately and you should neither act nor forebear from acting on account of the instruction.

Yours faithfully

Copies: Employer
Quantity surveyor
Consultants
Clerk of works

函件 127a

关于分包事宜，致承包商

此信函不适用于WCD98 合同

Letter 127a

To contractor, regarding sub-letting

This letter is not suitable for use with WCD 98

尊敬的先生：

现致函提请你方注意，合同第19.2.2条［使用 IFC98 或 MW98 合同时，替换为"第3.2条"，使用 GC/Works/1（1998）合同时，替换为"第62（1）条"］规定，禁止在不经我方同意的情况下将工程的任何部分分包，［如果使用 GC/Works/1（1998）合同，加上：］除非在签署合同之前雇主接受了分包建议书。

请注意，在我方同意之前，要求你方告知被建议的分包商名称和希望进行分包的工程部分。

你忠诚的

Dear Sir

I am writing to draw your attention to clause 19.2.2 [*substitute '3.2' when using IFC 98 or MW 98, or '62 (1)' when using GC/Works/1 (1998)*] of the contract which prohibits sub-letting any part of the Works without my consent [*add, if using GC/Works/1 (1998):*] unless the employer accepted a sub-letting proposal prior to the award of the contract.

Will you please note that, before giving my consent, I will require you to inform me of the names of the proposed sub-contractors and the parts of the Works you wish to sub-let.

Yours faithfully

函件 127b

关于分包事宜，致承包商

此信函仅适用于 WCD98 合同

Letter 127b

To contractor, regarding sub-letting

This letter is only suitable for use with WCD 98

尊敬的先生：

现致函提请你方注意，合同第 18. 2. 1 条和第 18. 2. 3 条规定，禁止在不经我方同意的情况下将工程的任何部分或设计工作的任何部分进行分包。

请注意，在我方同意之前，要求你方告知被建议的分包商名称和希望进行分包的工程部分或设计工作部分。如果我方同意按第 18. 2. 3 条将设计工作进行分包，此项决议丝毫不会影响你方履行合同第 2. 5 条义务。

你忠诚的

Dear Sir

I am writing to draw your attention to clauses 18. 2. 1 and 18. 2. 3 of the contract which prohibit sub-letting of any part of the Works or the design of any part of the Works respectively without my consent.

Will you please note that, before giving my consent, I will require you to inform me of the names of the proposed sub-contractors and the parts of the Works you wish to sub-let or of which you wish to sub-let the design. If I give my consent for the sub-letting of design under clause 18. 2. 3, such consent will not affect in any way your obligation under clause 2. 5.

Yours faithfully

函件 128

如果建筑师是雇主的员工，致承包商

Letter 128

To contractor, if architect is employee of the employer

尊敬的先生：

在工程尚未开始之前，借此机会致函向你方明确说明我方的职务，我方既是合同雇主的雇员，也是同一合同中的建筑师。

合同中规定了我方的职权范围，在合同规定条款之外，我方不具有制约雇主的总代理权。如果我方必须按合同条款做出决定，我方会在各方之间对所涉事项做出公平决断。如果我方要代表雇主致函给你方，我方会明确说明。

你忠诚的

Dear Sir

I am taking this opportunity of writing to you before work begins on site in order to make my position clear as employee of the employer named in the contract and also as the architect named in the same contract.

The extent of my authority is laid down in the contract and I have no general power of agency to bind the employer outside its express terms. If I have to make decisions under the contract terms, I will decide all such matters impartially between the parties. If I have cause to write to you on behalf of the employer, I will clearly so state.

Yours faithfully

第十章　工地施工活动

除非项目很小，在这一阶段通常要撰写很多信函。很多事务可用到标准信函，主要涉及当承包商没有投保或没有遵守指示、规定实施工程、延期、预定违约金、损失和/或支出费用、合同终止、裁决和仲裁进行工作时的处理程序。但在整个施工阶段还有很多事项也需要用到标准信函。

裁决和仲裁是专业性的话题。虽然你应该大体了解一下这些程序，但是即便雇主希望你能协助他委任的专人进行裁决和仲裁事务，你也不适合插手处理此类事件。

我方承认很多信函可以在任何时间发出，但是我方也努力把这些信函按某种合理的顺序进行排列。因此，在有关未能投保事项的信函之后，是进驻工地方面的信函，然后是各种关于指示的信函和工程延期的信函等等。

希望你方能够选择适合某种既定情况的恰当的信函，并加以修改使之符合你方的特定需要。

函件 129a

如果承包商没有投保，致承包商

此信函仅适用于 JCT98 合同和 WCD98 合同

Letter 129a

To contractor, if he fails to maintain insurance cover

This letter is only suitable for use with JCT 98 and WCD 98

尊敬的先生：

今晨我方与你方［填入姓名］进行了电话谈话，并确认你方不能提供保险单和保险费收据或其他文件证据，以证明你方进行了第 21.1.1.1/21.2.1/22A.1/22D.1［视情况取舍］条要求的保险。

鉴于保险的重要性，并且不影响你方［如果没能履行第 21.1.1.1 条，填入“根据合同第 20 条”］的责任，雇主打算立即按合同第 21.1.3/21.2.4/22A.2 条［使用 WCD98 合同要求时，替换为“21.2.3 条”］/22D.4 条［视情况取舍］行使权利。［涉及到第 21.1.1.1 条和第 21A.1 条，以及使用 WCD98 合同时的第 21.2.1 条，加上：］所有雇主所付的保险费用将从付给你方或将付给你方的款项中扣除[1]或者将作为你方的债务而得到偿付。［涉及到第 22D.1 条以及使用 JCT98 合同时的第 22.2.1 条，加上：］任何雇主支付的保险费不会包含在合同总金额的调整中。

你忠诚的

抄送：雇主

工料测量师

［1 雇主必须发出相应的扣除通知。］

Dear Sir

I refer to my telephone conversation with your [*insert name*] this morning and confirm that you are unable to produce the insurance policy, premium receipts or other documentary evidence that the insurance required by clause 21. 1. 1. 1/21. 2. 1/22A. 1/22D. 1 [*delete as appropriate*] is being maintained.

In view of the importance of the insurance and without prejudice to your liabilities [*if a clause 21. 1. 1. 1 failure, insert: 'under clause 20 of the conditions of contract'*], the employer is arranging to exercise his rights under clause 21. 1. 3/21. 2. 4/22A. 2 [*'21. 2. 3' when using WCD 98*] /22D. 4 [*delete as appropriate*] immediately. [*In respect of clauses 21. 1. 1. 1, 22A. 1 and, when using WCD 98, clause 21. 2. 1, add:*] Any sum or sums payable by him in respect of premiums will be deducted from any monies due or to become due to you[1] or will be recovered from you as a debt. [*In respect of clause 22D. 1 and, when using JCT 98, clause 21. 2. 1, add:*] Any sum or sums payable by him in respect of premiums will not be included in the adjustment of the contract sum.

Yours faithfully

Copies: Employer
Quantity surveyor

[1 *The employer must serve the appropriate withholding notices.*]

函件 129b

如果承包商没有投保，致承包商

此信函仅适用于 IFC98 合同

Letter 129b

To contractor, if he fails to maintain insurance cover

This letter is only suitable for use with IFC 98

尊敬的先生：

今晨我方与你方［填入姓名］进行了电话谈话，并确认你方不能提供保险单和保险费收据或其他文件证据，以证明你方进行了第 6.2.1/6.2.4/6.3A.1/6.3D.1［视情况取舍］条要求的保险。

鉴于保险的重要性，并且不影响你方［如果没能履行第 6.2.1 条，填入"根据合同条件第 6.1 条"］的责任，雇主打算立即按合同第 6.2.3/6.2.4/6.3A.2/6.3D.4 条［视情况取舍］行使权利。［涉及到第 6.2.1 条和第 6.3A.1 条，加上：］所有雇主所付的保险费用将从付给你方或将付给你方的款项中扣除[1]或将作为你方的债务而得到偿付。［关于第 6.2.4 条和第 6.3D.1 条，加上：］任何雇主支付的保险费不会包含在合同总金额的调整中。

你忠诚的

抄送：雇主

工料测量师

［1 雇主必须发出相应的扣除通知。］

Dear Sir

I refer to my telephone conversation with your [*insert name*] this morning and confirm that you are unable to produce the insurance policy, premium receipts or other documentary evidence that the insurance required by clause 6. 2. 1/6. 2. 4/6. 3A. 1/6. 3D. 1 [*delete as appropriate*] is being maintained.

In view of the importance of the insurance and without prejudice to your liabilities [*if a clause 6. 2. 1 failure, insert 'under clause 6. 1 of the conditions of contract'*], the employer is arranging to exercise his rights under clause 6. 2. 3/6. 2. 4/6. 3A. 2/6. 3D. 4 [*delete as appropriate*] immediately. [*In respect of clauses 6. 2. 1 and 6. 3A. 1 add:*] Any sum or sums payable by him in respect of premiums will be deducted from any monies due or to become due to you[1] or will be recovered from you as a debt. [*In respect of clauses 6. 2. 4 and 6. 3D. 1 add:*] Any sum or sums payable by him in respect of premiums will not be included in the adjustment of the contract sum.

Yours faithfully

Copies: Employer
Quantity surveyor

[1 *The employer must serve the appropriate withholding notices.*]

函件 129c

如果承包商没有投保，致承包商

此信函仅适用于 MW98 合同

Letter 129c

To contractor, if he fails to maintain insurance cover

This letter is only suitable for use with MW 98

尊敬的先生:

今晨我方与你方［填入姓名］进行了电话谈话，并确认你方不能提供雇主按照合同条件第 6.4 条合理要求的证据来证明你方进行了第 6.1/6.2/6.3A 条［视情况取舍］要求的保险，并已经生效。

这明显是你方违约。鉴于保险的重要性，并且不影响你方按照第 6.1/6.2/6.3A 条［视情况取舍］规定的责任，雇主正在准备以你方名义投保。所有雇主应付的保险费用将从付给你方或将付给你方的款项中扣除[1]或将作为你方的债务而得到偿付。

你忠诚的

抄送：雇主

工料测量师［如果已经委任］

［1 雇主必须发出相应的扣除通知。］

Dear Sir

I refer to my telephone conversation with your [*insert name*] this morning and confirm that you are unable to produce such evidence as the employer may reasonably require in accordance with clause 6. 4 of the conditions of contract that the insurances referred to in clause 6. 1/6. 2/6. 3A [*delete as appropriate*] have been taken out and are in force.

This is a clear breach of contract on your part and, in view of the importance of the insurance and without prejudice to your liabilities under clause 6. 1/6. 2/6. 3A [*delete as appropriate*], the employer is arranging to take out the appropriate insurances on your behalf. Any sum or sums payable by him in respect of premiums will be deducted from any monies due or to become due to you[1] or will be recovered from you as a debt.

Yours faithfully

Copies: Employer
Quantity surveyor (if appointed)

[1 *The employer must serve the appropriate withholding notices.*]

函件 129d

如果承包商没有投保，致承包商

此信函仅适用于 GC/Works/1（1998）合同

Letter 129d

To contractor, if he fails to maintain insurance cover

This letter is only suitable for use with GC/Works/1 (1998)

尊敬的先生：

今晨我方与你方［填入姓名］进行了电话谈话，并确认你方不能提供保险单和保险费收据或其他文件证据，以证明你方进行了第 8 条要求的保险。

鉴于保险的重要性，并且不影响合同条件第 8 条规定的你方的责任，雇主打算立即按合同第 8(4)条行使权利。投保的费用将会从按合同付给你方的预付款中扣除[1]。

你忠诚的

抄送：雇主
　　工料测量师［如果已经委任］

［1 雇主必须发出相应的扣除通知。］

Dear Sir

I refer to my telephone conversation with your [*insert name*] this morning and confirm that you are unable to produce the insurance policy, premium receipts or other documentary evidence that the insurance required by clause 8 is being maintained.

In view of the importance of the insurance and without prejudice to your liabilities under clause 8 of the conditions of contract, the employer is arranging to exercise his rights under clause 8(4). The cost of effecting the appropriate insurance cover will be deducted from any advance payment due to you[1] under the contract.

Yours faithfully

Copies: Employer
　　Quantity surveyor (if appointed)

[1 *The employer must serve the appropriate withholding notices.*]

函件 130a

如果承包商没有投保，致委托人

此信函仅适用于 JCT98 合同或 WCD98 合同

Letter 130a

To client, if contractor fails to maintain insurance cover

This letter is only suitable for use with JCT 98 or WCD 98

尊敬的先生：

承包商不能提供证据证明其已经按合同条件第 21. 1. 1. 1/21. 2. 1/22A. 1/22D. 1［视情况取舍］条投了保险。

鉴于保险的重要性，我方已经代表你方按合同第 21. 1. 3/21. 2. 4/22A. 2［使用 WCD98 合同时，替换为“21. 2. 3 条”］/22D. 4 条［视情况取舍］采取了行动。我方已指示你方的保险经纪人办理保险，今日生效。［涉及到第 21. 1. 1. 1 条和第 21A. 1 条，以及使用 WCD98 合同时的第 21. 2. 1 条，加上：］你方有权从下次向承包商的付款中扣除保险费金额或者把此金额作为对方的债务而取得偿付。如果你方希望扣除此金额，请记住向承包商发出有效的书面扣除通知。［涉及到第 22D. 1 条以及使用 JCT98 合同时的第 21. 2. 1 条，加上：］正常情况下，你方支付的保险费不会包含在合同总金额的调整中。

你忠诚的

Dear Sir

The contractor is unable to provide evidence that he is maintaining the insurance cover required by clause 21. 1. 1. 1/21. 2. 1/22A. 1/22D. 1 [*delete as appropriate*] of the conditions of contract.

In view of the importance of the insurance, I have taken action on your behalf, under clause 21. 1. 3/21. 2. 4/22A. 2 [*'21. 2. 3' when using WCD 98*] /22D. 4 [*delete as appropriate*] and instructed your broker to provide the necessary cover effective from today. [*In respect of clauses 21. 1. 1. 1, 22A. 1 and, when using WCD 98, clause 21. 2. 1, add*:] You are entitled to deduct the amount of the premium from your next payment to the contractor or, alternatively, you may wish to recover it as a debt. If you wish to deduct, remember to give the contractor an effective written withholding notice. [*In respect of clause 22D. 1 and, when using JCT 98, clause 21. 2. 1, add*:] Naturally, the amount of your premium will not be included in the adjustment of the contract sum.

Yours faithfully

函件 130b

如果承包商没有投保，致委托人

此信函仅适用于 IFC98 合同

Letter 130b

To client, if contractor fails to maintain insurance cover

This letter is only suitable for use with IFC 98

尊敬的先生：

承包商不能提供证据证明其已经按合同条件第 6.2.1/6.2.4/6.3A.1/6.3D.1 [视情况取舍] 条投了保险。

鉴于保险的重要性，我方已经代表你方按合同第 6.2.3/6.2.4/6.3A.2/6.3D.4 [视情况取舍] 采取了行动。我方已指示你方的保险经纪人办理保险，今日生效。[涉及到第 6.2.1 条和第 6.3A.1 条，加上：] 你方有权从下次向承包商的付款中扣除保险费金额或者把此金额作为对方的债务而取得偿付。如果你方希望扣除此金额，请记住向承包商发出有效的书面扣除通知。[关于第 6.2.4 和第 6.3D.1 条，加上：] 正常情况下，你方支付的保险费不会包含在合同总金额的调整中。

你忠诚的

Dear Sir

The contractor is unable to provide evidence that he is maintaining the insurance cover required by clause 6.2.1/6.2.4/6.3A.1/6.3D.1 [*delete as appropriate*] of the conditions of contract.

In view of the importance of the insurance, I have taken action on your behalf, under clause 6.2.3/6.2.4/6.3A.2/6.3D.4 [*delete as appropriate*], and instructed your broker to provide the necessary cover effective from today. [*In respect of clauses 6.2.1 and 6.3A.1 add*:] You are entitled to deduct the amount of the premium from your next payment to the contractor or, alternatively, you may wish to recover it as a debt. If you wish to deduct, remember to give the contractor an effective written withholding notice. [*In respect of clauses 6.2.4 and 6.3D.1 add*:] Naturally, the amount of your premium will not be included in the adjustment of the contract sum.

Yours faithfully

函件 130c

如果承包商没有投保，致委托人

此信函仅适用于 MW98 合同

Letter 130c

To client, if contractor fails to maintain insurance cover

This letter is only suitable for use with MW 98

尊敬的先生：

承包商不能按合同条件第 6.4 条规定，提供证据证明其已经进行了合同条件第 6.1/6.2/6.3A 条［视情况取舍］规定的保险并已经生效。

鉴于保险的重要性，我方已经代表你方采取了行动。我方指示你方的保险经纪人办理保险，今日生效。尽管合同没有明确条款说明你方可以对承包商的违约采取行动，但我方认为，此时你方有权从下次向承包商的付款中扣除保险费金额。如果你方希望扣除此金额，请记住向承包商发出有效的书面扣除通知。

你忠诚的

Dear Sir

The contractor is unable to provide evidence in accordance with clause 6.4 of the conditions of contract that he is maintaining the insurance cover required by clause 6.1/6.2/6.3A [*delete as appropriate*] of the conditions of contract.

In view of the importance of the insurance, I have taken action on your behalf and instructed your broker to provide the necessary cover effective from today. Although the terms of the contract make no express provision for you to act on the contractor's default, it is my opinion that, in this instance, you are entitled to deduct the cost of the premium from your next payment to the contractor. If you wish to deduct, remember to give the contractor an effective written withholding notice.

A copy of my letter to the contractor, dated [*insert date*], is enclosed for your information.

Yours faithfully

函件 130d

如果承包商没有投保，致委托人

此信函仅适用于 GC/Works/1（1998）合同

Letter 130d

To client, if contractor fails to maintain insurance cover

This letter is only suitable for use with GC/Works/1 (1998)

尊敬的先生：

承包商不能提供证据证明他已经进行了合同条件第 8 条规定的保险。

鉴于保险的重要性，我方已经代表你方按合同第 8(7)条采取了行动。我方已指示你方的保险经纪人办理保险，今日生效。你方有权从付给承包商的预付款中扣除保险费用。如果你方希望扣除此金额，请记住向承包商发出有效的书面扣除通知。

你忠诚的

Dear Sir

The contractor is unable to provide evidence that he is maintaining the insurance cover required by clause 8 of the conditions of contract.

In view of the importance of the insurance, I have taken action on your behalf, under clause 8(7), and instructed your broker to provide the necessary cover effective from today. You are entitled to deduct the cost of so insuring from any advance payment due to the contractor under the contract. If you wish to deduct, remember to give the contractor an effective written withholding notice.

Yours faithfully

函件 131

如果承包商没有投保，致委托人的保险经纪人

Letter 131

To client's insurance broker, if contractor fails to maintain insurance cover

尊敬的先生：

我方的委托人［填入姓名］已经就以上项目向你方进行了咨询，现关于此项目致函于你方。我方会单独致函我方的委托人，并附上此函副本。

承包商没有按照合同条件第［使用 JCT98 合同时，填入第 21. 1. 1. 1/22A. 1/21. 2. 1/22D. 1 条之一；使用 IFC98 合同时，填入第 6. 2. 1/6. 3A. 1/6. 2. 4/6. 3D. 1 条之一；使用 MW98 合同时，填入 6. 1/6. 2/6. 3A 条之一；使用 GC/Works/1（1998）合同时，填入第 8 条］进行投保。因此，我方的委托人有必要自行投保。随函附上合同相关部分的摘录。请立即组织投保。我方的委托人将支付保险费，但请你方告知我方具体金额，以便从承包商处获得偿付。

你忠诚的

抄送：雇主

工料测量师

Dear Sir

I am writing to you in connection with the above project about which my client [*insert name*] has already consulted you. I am writing separately to my client and sending him a copy of this letter.

The contractor has failed to maintain/take out [*delete as appropriate*] insurance cover as required by clause [*insert one of 21. 1. 1. 1/22A. 1/21. 2. 1/22D. 1 when using JCT 98*, *insert one of 6. 2. 1/6. 3A. 1/6. 2. 4/6. 3D. 1 when using IFC 98*, *insert 6. 1/6. 2/6. 3A when using MW 98 or insert 8 when using GC/Works/1 (1998)*] of the conditions of contract and it is, therefore, necessary that my client takes out such insurance himself. The relevant extract from the contract is enclosed. Please arrange insurance cover immediately. The premium will be paid by my client, but you should let me know the amount so that it can be recovered from the contractor.

Yours faithfully

Copies: Employer
Quantity surveyor

函件 132

致承包商，确认进驻工地

Letter 132

To contractor, confirming possession of the site

尊敬的先生：

我方现确认以上工地对你方开放，你方可以按合同规定于［填入日期］进驻工地。

你忠诚的

抄送：雇主

Dear Sir

I confirm that the above site was available for you to take possession on the [*insert date*], which is the date stated in the contract.

Yours faithfully

Copy: Employer

函件 133

雇主致承包商的函件草稿，推迟进驻工地

此信函不适用于 MW98 或 GC/Works/1（1998）合同

Letter 133

Draft letter from employer to contractor, deferring possession of the site

This letter is not suitable for use with MW 98 or GC/Works/1 (1998)

尊敬的先生：

[如果知道推迟的期限：]

按照合同条件第 23.1.2 条［使用 IFC98 合同时，替换为“第 2.2 条”］，现通知你方，工地移交将推迟［填入不多于 6 周的时间段］。你方可以于［填入日期］进驻工地。

[如果不知道推迟的期限：]

按照合同条件第 23.1.2 条［使用 IFC98 合同时，替换为“第 2.2 条”］，现通知你方，工地移交将推迟不超过［填入附件中规定的期限］的时间。确定进驻工地具体时间后，我方将再次通知你方。

你忠诚的

抄送：建筑师

工料测量师

顾问

工程监督员

Dear Sir

[*If length of deferment is known*:]

I hereby give you notice in accordance with clause 23. 1. 2 [*substitute* '*2. 2*' *when using IFC 98*] of the conditions of contract that I defer giving possession of the site for [*insert period which should not exceed 6 weeks*]. You may take possession of the site on [*insert date*].

[*if length of deferment is not known*:]

I hereby give you notice in accordance with clause 23. 1. 2 [*substitute* '*2. 2*' *when using IFC 98*] of the conditions of contract that I defer giving possession of the site for a period which will not exceed [*insert period named in the appendix*]. I will write to you again when I have a definite date for possession.

Yours faithfully

Copies: Architect
Quantity surveyor
Consultants
Clerk of works

函件 134

如果《承包商报表》不完善，致承包商

此信函仅适用于 JCT98 合同

Letter 134

To contractor, if Contractor's Statement is deficient

This letter is only suitable for use with JCT 98

尊敬的先生：

对你方于［填入日期］的来信深表谢意，你方按附件说明［简要描述附件中对此工作的说明］，就相关的规定实施工程附上了《承包商报表》。

我方认为此报表不足以解释你方的建议书，因为［填入不完善的原因］。

请将此函作为依据合同条件第 42.5 条的正式通知，请你方修改此报表，以便使之尽量完善。

［视情况，加上：］

按第 42.6 条，此信函也作为书面通知。

你忠诚的

Dear Sir

Thank you for your letter of the [*insert date*] with which you enclosed a copy of your Contractor's Statement in respect of the performance specified work identified in the appendix as [*insert brief description of the work as noted in the appendix*].

It is my opinion that such Statement is deficient adequately to explain your proposals, because [*insert reason why it is deficient*].

Take this letter as written notice under clause 42.5 of the conditions of contract requiring you to amend such Statement so that it will no longer be deficient as noted above.

[*Add, if appropriate*:]

This letter also serves as written notice under clause 42.6.

Yours faithfully

函件 135

致承包商，要求合同总额分析书

此信函仅适用于 JCT98 合同

Letter 135

To contractor, requiring the Analysis

This letter is only suitable for use with JCT 98

尊敬的先生：

合同清单中，没有提供关于规定实施工程的合同金额分析书。因此按合同条款第 42. 13 条，现要求你方于收到此信日期 14 天内，提供此分析书。

你忠诚的

Dear Sir

No analysis of the portion of the contract sum which relates to performance specified work is provided in the contract bills. Therefore, in accordance with clause 42. 13 of the conditions of contract, I require you to provide such an analysis ('Analysis') within 14 days of the date of this letter.

Yours faithfully

函件 136

如果建筑师发现了《承包商报表》中的矛盾之处，致承包商

此信函仅适用于 JCT98 合同

Letter 136

To contractor, if architect finds discrepancy in Contractor's Statement

This letter is only suitable for use with JCT 98

尊敬的先生：

依据合同条件第 2. 4. 2 条，现提请你方注意以下我方发现的矛盾之处。

[列出不符之处，说明细节]

请改正报表中的矛盾之处，并书面通知我方你方进行的改动。注意，这些改动的费用不应由雇主支付。

你忠诚的

Dear Sir

In accordance with clause 2. 4. 2 of the conditions of contract, I bring to your attention the following discrepancies which I have discovered:

[*List, giving precise details*]

Please correct the Statement to remove the discrepancy and inform me in writing of the correction you have made. Note that such correction must be at no cost to the employer.

Yours faithfully

函件 137

如果建筑师发现《承包商报表》与法律要求有分歧，致承包商

此函件仅适用于 JCT98 合同

Letter 137

To contractor, if architect finds divergence between the Contractor's Statement and statutory requirements

This letter is only suitable for use with JCT 98

尊敬的先生：

我方发现你方的报表与法律要求有分歧，依据合同条款第 6.1.6 条，现向你方发出如下书面通知：

［填入分歧细节］

请你方将消除分歧的建议修正稿以书面形式寄给我方，以便我方可以立即发出指示。［如果第 6.1.7 条或第 42.15 条不适用，加上：］你方遵循指示进行修改的相关费用不由雇主支付。

你忠诚的

Dear Sir

I have found what appears to be a divergence between your Contractor's Statement and statutory requirements and, in accordance with clause 6.1.6 of the conditions of contract, I hereby give you written notice as follows:

[*Insert details of the divergence*]

Please inform me in writing of your proposed amendment to remove the divergence so that I can issue instructions without delay. [*If clauses 6.1.7 or 42.15 are not applicable, add*:] Your compliance with such instructions will be without cost to the employer.

Yours faithfully

函件 138

如果建筑师发现《雇主要求》中的矛盾之处，致承包商

此信函仅适用于 WCD98 合同

Letter 138

To contractor, if architect finds discrepancy within the Employer's Requirements

This letter is only suitable for use with WCD 98

尊敬的先生：

我方注意到《雇主要求》中的以下矛盾之处：[填入细节]。

[然后：]

你方《承包商建议书》的［填入参照条款］就此作出了规定。依据合同条件第 2.4.1 条，这些建议为有效。合同总额不需要进行任何调整。

[或：]

你方《承包商建议书》没有涉及此项，因而我方欣然希望你方遵循合同条件第 2.4.1 条，并书面告知你方处理矛盾后的建议修正稿，以便取得雇主同意或由其决定怎样处理矛盾之处。

你忠诚的

Dear Sir

I note the following discrepancy within the Employer's Requirements [*insert details*].

[*Then either*:]

Your Contractor's Proposals [*insert reference to item*] deal with the matter and, therefore, they prevail in accordance with clause 2. 4. 1 of the conditions of contract. There will be no adjustment to the contract sum.

[*Or*:]

Your Contractor's Proposals do not deal with the matter and, therefore, I should be pleased if you would comply with clause 2. 4. 1 of the conditions of contract and inform me in writing of your proposed amendment to deal with the discrepancy so that the employer may either agree or decide how to deal with the discrepancy.

Yours faithfully

函件 139

如果建筑师在《承包商建议书》中发现矛盾之处，致承包商

此信函仅适用于 WCD98 合同

Letter 139

To contractor, if architect finds discrepancy within the Contractor's Proposals

This letter is only suitable for use with WCD 98

尊敬的先生：

我方注意到《承包商建议书》中的以下矛盾之处：［填入细节］。

依据合同条件第 2.4.2 条，我方欣然希望收到你方消除矛盾后的建议修正稿，以便雇主决定选用矛盾条款中的一条或接受你方的建议修正稿。你方遵循雇主决议而引起的费用不由雇主支付。

你忠诚的

Dear Sir

I note the following discrepancy within the Contractor's Proposals [*insert details*].

In accordance with clause 2. 4. 2 of the conditions of contract I should be pleased to receive your proposed amendment to remove the discrepancy so that the employer may either decide between the discrepant items or may accept your proposed amendment. Your compliance with such decision or acceptance will be without cost to the employer.

Yours faithfully

函件 140

致承包商，要求对方同意在第 19.3 条的列表中添加人员

此信函仅适用于 JCT98 合同

Letter 140

To contractor, requesting consent to the addition of persons to the list under clause 19.3

This letter is only suitable for use with JCT 98

尊敬的先生：

我方谨代表我方的委托人［填入姓名］，依据合同条件第 19.3.2.1 条规定，现正式要求你方同意在合同参考［填入参考事项］清单中的指定人员名单上添加［填入姓名］［填入地址］，以便进行［填入合同清单中给出的工作描述］。

你忠诚的

Dear Sir

On behalf of the employer, [*insert name*], I hereby formally request your consent, as required under clause 19.3.2.1 of the conditions of contract, to the addition of [*insert name*] of [*insert address*] to the list of persons named in the contract bills reference [*insert reference*] to carry out [*insert description of the work as given in the contract bills*].

Yours faithfully

函件 141

如果承包商在未经建筑师同意的情况下对工程进行分包，致承包商

此信函不适用于 GC/Works/1 （1998） 合同

Letter 141

To contractor, if he sub-lets without architect's consent

This letter is not suitable for use with GC/Works/1 (1998)

尊敬的先生：

我方得知你方已将［填入分包的工程内容］分包给［填入分包商名称］。

因为未经我方的同意，你方的行为属违约行为，必须停止进行。［加上，除非使用 MW98 合同：］我方并不打算建议雇主行使第 27. 2. 1. 4 条规定的权利［使用 IFC98 合同时，替换为“第 7. 2. 1(d)条”］。

你忠诚的

Dear Sir

I am informed that you have sub-let [*insert part of the Works sub-let*] to [*insert name of sub-contractor*].

Because I have not given my consent, your action is a breach of contract and it must cease forthwith. Please confirm, by return, that you will comply with this letter. [*Add, except when using MW 98:*] I have no wish to advise the employer to use his powers under clause 27. 2. 1. 4 [*substitute '7. 2. 1(d)' when using IFC 98*].

Yours faithfully

函件 142a

关于雇主授权许可，致承包商

此信函不适用于 MW98 合同或 GC/Works/1（1998）合同

Letter 142a

To contractor, regarding employer's licensees

This letter is not suitable for use with MW 98 or GC/Works/1 (1998)

尊敬的先生：

雇主正急于安排［填入工程的性质］于［填入日期］开工。因为此项工作不属于合同的一部分，合同文件也没有规定雇主自行进行此工作。按照其指示，现要求你方同意由雇主直接雇佣的人员进行此项工作。

此要求是依据第 29. 2 条提出的［使用 IFC98 合同时，替换为“第 3. 11 条”］。

你忠诚的

Dear Sir

The employer is anxious to arrange [*insert the nature of the work*] commencing on [*insert date*]. Because this work does not form part of the contract and the contract documents do not provide for the employer to carry out the work himself, he has instructed me to request your consent to the carrying out of such work by persons employed directly by the employer.

This request is made in accordance with the provisions of clause 29. 2 [*substitute '3. 11' when using IFC 98*].

Yours faithfully

函件 142b

关于雇主授权许可，致承包商

此信函只适用于 GC/Works/1 （1998） 合同

Letter 142b

To contractor, regarding employer's licensees

This letter is only suitable for use with GC/Works/1 (1998)

尊敬的先生：

雇主指示我方通知你方，依据合同条件第65(1)条，雇主打算任用由其直接雇佣的承包商来进行［填入工程性质］。此工程将于［填入日期］开工。

如果你方认为此工程不应该按以上所述方式进行，请立即通知我方。

你忠诚的

Dear Sir

The employer instructs me to inform you that, in accordance with clause 65(1) of the conditions of contract, he intends to carry out [*insert the nature of the work*] using his own directly employed contractors. Work will commence on [*insert date*].

Please inform me immediately if you consider that the work should not proceed as indicated.

Yours faithfully

函件 143

致委托人，随函附上进度报告

Letter 143

To client, enclosing report on progress

尊敬的先生:

现欣然附上详细说明了项目进度的编号为［填入编号］的报告，如果你方需要我方对任何部分加以进一步详述，我方将欣然照办。

［如果此次为第一次寄去进度报告，加上:］

你方将每隔［填入间隔时间］收到此种形式的进度报告。如果以后报告中有任何方面需要我方作详细解释，请通知我方。

你忠诚的

Dear Sir

I have pleasure in enclosing my report number [*insert number*] giving details of the progress of this project. If you wish me to enlarge upon any part of the report, I shall be happy to do so.

[*If this is the first time a progress report is being sent, add*:]

You will receive further progress reports in this form at [*insert period*] intervals. If there is any aspect of the report on which you wish me to elaborate in the future, please let me know.

Yours faithfully

函件 144

致相关人员，随函附上会议记录摘要

Letter 144

To persons affected, enclosing extract of minutes

尊敬的先生：

[填入主要参加人的姓名] 于 [填入日期] 日进行了会议，作出如下的决议：

[填入会议记录，包括参考内容条目，然后：]

欣然希望你方能采取相应的行动。

[或：]

欣然希望收到你方的意见。

[或：]

请将会议记录备案。

你忠诚的

Dear Sir

At a meeting held on the [*insert date*] between [*insert names of principal participants*], it was resolved:

[*Insert minute, including reference number, then either*:]

I should be pleased if you would take the appropriate action.

[*Or*:]

I should be pleased to receive your comments.

[*Or*:]

Please note the minute in your records.

Yours faithfully

函件 145

如果建筑师不同意其中的内容，致会议记录整理人

Letter 145

To originator of minutes, if architect disagrees with contents

尊敬的先生：

我方仔细阅读了今天收到的［填入日期］的会议记录，以下为我方的意见：

［填入意见列表］

请在下次会议上公布这些意见，并填入会议记录适当的位置。

你忠诚的

抄送：所有到会者和寄发的会议记录人名单上的人员

Dear Sir

I have considered the minutes of the meeting held on the [*insert date*] which I received today and I have the following comments:

[*Insert list of comments*]

Please arrange to have these comments published at the next meeting and inserted in the appropriate place in the minutes.

Yours faithfully

Copies: All present at the meeting and all included in the circulation list

函件 146a

如果不同意前任建筑师的决定，致委托人

此信函不适用于 WCD98 合同

Letter 146a

To client, if disagreeing with former architect's decisions

This letter is not suitable for use with WCD 98

尊敬的先生：

我方接受委任后，审查了所有移交给我方的工程图纸和文件，勘查了工地，并与承包商和工程监督员进行了谈话。

我方不同意前任建筑师做出的下列决定：

［列出不同意的所有事项］

如果你方有时间，我方建议面议如何处理这些事项。［如果使用 JCT98 合同、IFC98 合同或 MW98 合同时，加上：］合同第 3 条规定，我方不能无视或推翻之前被雇佣管理此工程合同的建筑师发出的证明书或指示。

你忠诚的

Dear Sir

I have examined all the drawings and documents handed to me on appointment and I have visited site and spoken to the contractor and the clerk of works.

I find myself unable to agree with the following decisions of the former architect:

[*List all matters with which you disagree*]

When you have had time to study these matters, I suggest we should meet to discuss ways of dealing with them. [*Add, if using JCT 98, IFC 98 or MW 98*:] Article 3 of the contract prevents me from disregarding or overruling any certificate or instruction given by the architect previously engaged to administer this contract.

Yours faithfully

函件 146b

如果不同意前任建筑师的决定，致委托人

此信函仅适用于 WCD98 合同

Letter 146b

To contractor client, if disagreeing with former architect's decisions

This letter is only suitable for use with WCD 98

尊敬的先生：

我方接受委任后，审查了所有移交给我方的工程图纸和文件，并勘查了工地。我方注意到《雇主要求》中包含了大量的雇主顾问完成的详细图纸。其中有不少事项已引起我方的担忧，但在目前的情况下，可能已经无力改变现状了。

如果你方有时间，我方建议面议如何处理这些事项。

你忠诚的

Dear Sir

I have examined all the drawings and documents handed to me on appointment and I have visited site. I note that the Employer's Requirements contain a substantial number of relatively detailed drawings produced by the employer's consultant. There are a number of matters which give me cause for concern, but presumably there is little which can be done about it at this stage:

[*List all matters with which you disagree*]

When you have had time to study these matters, I suggest we should meet to discuss ways of dealing with them.

Yours faithfully

函件 147

致承包商，收到总进度计划

Letter 147

To contractor, on receipt of master programme

尊敬的先生：

你方于［填入日期］的来函并附件总进度计划收悉，深表谢意。我方的意见如下：

［以问题的方式列出意见，而不是以指示方式来改变总进度计划的具体内容］

欣然希望你方能按我方的意见重新考虑此计划，但我方的意见或无任何意见不应该被认为是对此计划的部分或全部的批准。你方的责任是按工程图纸和工程量清单［或工程详细说明］规定制定出工程的组织模式和施工方法，并为具体活动分配时间。

你方的总进度计划仅被作为你方意图的表述。只有经过我方的决定，此计划方可作为将来要求工程延期或损失和/或费用[1]的证据。

你忠诚的

［1 使用 MW98 合同时，省略划线部分。］

Dear Sir

Thank you for your letter of the [*insert date*] with which you enclosed two copies of your master programme. My comments are as follows:

[*List comments as questions, not as instructions to change particular parts of the programme*]

I should be pleased if you would reconsider your programme in the light of my comments, but you must not take such comments or lack of comment to indicate approval to the programme in part or in whole. The organisation and method of working and the times allocated to particular activities are your responsibility to carry out within the constraints laid down by the drawings and bills of quantities [*or specification*].

Your master programme is only received as an indication of your intentions. The use of the programme as evidence in any future claim for extension of time or loss and/or expense[1] is subject to my discretion.

Yours faithfully

[[1] *Omit this phrase when using MW 98*]

函件 148

如果要求建筑师检查放线，致承包商

Letter 148

To contractor, if architect asked to check setting out

尊敬的先生：

你方于［填入日期］的来函收悉，深表谢意。

合同规定[1]放线是你方的责任。我方认为你方已经获得了所有必要的尺寸和信息以确保你方履行责任。任何我方决定进行的检查都不能减少你方的义务。

你忠诚的

［1 使用和 MW98 合同时，省略划线部分。］

Dear Sir

Thank you for your letter of the [*insert date*].

The setting out is your responsibility under the contract[1]. In my opinion, you have been provided with all necessary dimensions and information to enable you to discharge that responsibility. Any inspection I may decide to carry out will not remove any of your obligations.

Yours faithfully

[1 *Omit this phrase when using MW 98*]

函件 149

致承包商，要求提供合理证据证明已经向指定分包商付款

此信函仅适用于 JCT98 合同

Letter 149

To contractor, requiring reasonable proof that the nominated sub-contractors have been paid

This letter is only suitable for use with JCT 98

尊敬的先生：

我方随函附上于［填入日期］签发的［填入编号］中期证明书，以及关于每位指定分包商的应付金额的［填入编号］指示。

请注意，按第 35. 13. 3 条，你方必须在下次中期证明书发出之前提供合理的证据，以证明你方已经按 NSC/C 分包合同的规定按期向分包商支付了相应的合同金额。

你忠诚的

Dear Sir

I enclose my interim certificate number [*insert number*] dated [*insert date*] together with direction number [*insert number*] in respect of the amounts included for each nominated sub-contractor.

Please note that, in accordance with clause 35. 13. 3 you must provide reasonable proof, before the issue of the next interim certificate, that such sub-contract amounts have been duly discharged in accordance with sub-contract NSC/C.

Yours faithfully

函件 150

如果承包商不能提供合理证据证明已经向分包商付款，致雇主

此信函仅适用于 JCT98 合同

Letter 150

To employer, if contractor fails to provide reasonable proof of payment to sub-contractors

This letter is only suitable for use with JCT 98

尊敬的先生：

我方依据合同条件第35.13.5.2条，特此证明承包商不能提供合理的证据，以证明其已按我方于［填入日期］签发的［填入编号］证明书中的金额［填入金额］向［填入指定分包商的名称］付款。

你忠诚的

抄送：指定分包商
工料测量师
承包商

Dear Sir

In accordance with clause 35.13.5.2 of the conditions of contract, I hereby certify that the contractor has failed to provide me with reasonable proof that he has discharged the amount of [*insert amount*] included in my certificate number [*insert number*] dated [*insert date*] in respect of [*insert the name or names of the nominated sub-contractors concerned*].

Yours faithfully

Copies: Nominated sub-contractor
Quantity surveyor
Contractor

函件 151

如果发出了合同第 35. 13. 5. 1 条的证明书，致委托人

此信函仅适用于 JCT98 合同

Letter 151

To client, if clause 35. 13. 5. 1 certificate issued

This letter is only suitable for use with JCT 98

尊敬的先生：

现附上我方于［填入日期］按照第 35. 13. 5. 1 条签发的证明书。

依据合同第 35. 13. 5. 2 条，你方应该在不迟于下次中期证明书发出前 14 天，直接向［填入指定分包商的名称］支付［填入金额］，你有权按合同减少应付承包商的金额。

你忠诚的

Dear Sir

I enclose my certificate dated [*insert date*] issued in accordance with clause 35. 13. 5. 1.

In accordance with clause 35. 13. 5. 2 you should pay the sum of [*insert amount*] directly to [*insert name of nominated sub-contractor*] no later than 14 days from the date of my next interim certificate. You are entitled to reduce the amount payable to the contractor accordingly.

Yours faithfully

函件 152

致承包商，确认按第 13A 条接受报价

此信函仅适用于 JCT98 合同

Letter 152

To contractor, confirming acceptance under clause 13A

This letter is only suitable for use with JCT 98

尊敬的先生：

谨提及你方于［填入日期］寄来的关于合同条件第 13A 条的报价单。依据合同第 13A. 3. 2 条，现确认下列事项：

1. 雇主已经接受了你方报价。
2. 你方应该负责进行与报价相关的改动。
3. 合同总额应该通过增加/删除［填入金额］进行调整。此金额兼顾到第 13A. 2. 3 条和 13A. 2. 4 条所指的金额。
4. 修改后的竣工日期为［填入日期］。［视情况加上：］修改后的［填入指定分包商的名称］的竣工期限为［填入期限］。
5. 你方必须接受 NSC/C 报价第 3. 3A 条，此条已经包含在上述报价单中。

你忠诚的

Dear Sir

I refer to your quotation, pursuant to clause 13A of the conditions of contract, dated [*insert date*]. I write in accordance with clause 13A. 3. 2 of the contract and confirm the following:

1. The employer has accepted your quotation.
2. You are to carry out the variation to which the quotation relates.
3. The contract sum is to be adjusted by the addition/omission [*delete as appropriate*] of [*insert amount*] which takes account of amounts to which clauses 13A. 2. 3 and 13A. 2. 4 refer.
4. The revised date for completion is [*insert date*]. [*Add, if relevant*:] The revised period for completion of [*insert name of nominated sub-contract*] is [*insert period*].
5. You must accept any clause 3. 3A of NSC/C quotation which is included in the quotation for which this letter has been issued.

Yours faithfully

函件 153

关于价格报表中的要求，致工料测量师

此信函仅适用于 JCT98 合同或 IFC98 合同

Letter 153

To quantity surveyor, regarding a requirement in a priced statement

This letter is only suitable for use with JCT 98 or IFC 98

尊敬的先生：

你方于［填入日期］的来信收悉，深表谢意。信中列出了你方认为承包商基于合同第 A7.1.1 条（损失和/或费用）/A7.1.2 条（工程延期）［视情况取舍］的要求应该/不应该［视情况取舍］被接受的理由。我方清楚你方是依据第 A7.1 条的条款与我方进行商谈的。

认真考虑此事后，我方认为承包商的要求应该/不应该［视情况取舍］被接受。

你忠诚的

Dear Sir

Thank you for your letter dated [*insert date*] setting out your reasons for believing that the contractor's requirement under clause A7. 1. 1 (loss and/or expense) /A7. 1. 2 (extension of time) [*delete as appropriate*] should/should not [*delete as appropriate*] be accepted. I understand that you are consulting me under the provisions of clause A7. 1.

Having given the matter careful thought, I consider that the contractor's requirement should/should not [*delete as appropriate*] be accepted.

Yours faithfully

函件 154

如果意见不一致，工料测量师没有回应承包商的估价，致工料测量师

此信函仅适用于 JCT98 合同或 IFC98 合同

Letter 154

To quantity surveyor, if he fails to respond to contractor's valuation where there is disagreement

This letter is only suitable for use with JCT 98 or IFC 98

尊敬的先生：

我方从承包商处得知，他方已经于［填入日期］按合同第 30.1.2.2 条［使用 IFC98 合同时，替换为“第 4.2(c)条”］向你方提交了申请书。其中标出了他方基于合同第 30.2 条［使用 IFC98 合同时，替换为“第 4.2.1 条和第 4.2.2 条”］认为合理的总估价。但是他方没有收到你方的回信。

你方清楚进行中期评估是你方的义务，而且你方也履行了这一义务。但当意见不一致时，你方进一步的职责就是要回应承包商，为其提供一份说明，详细程度应该如同他方申请书一样，讲明你们双方意见不统一之处。承包商抱怨你方并未这样做。

或者你方可告知我方细节，以便我方回函给承包商。尽管我方不清楚承包商将采取什么补救措施，但如果你方不提供这份说明以讲清双方意见不同之处，便可能会出现违约的麻烦。

你忠诚的

Dear Sir

The contractor informs me that he submitted an application to you dated [*insert date*] under the provisions of clause 30. 1. 2. 2 [*substitute* '*4. 2(c)*' *when using IFC 98*] setting out what he considered to be the amount of gross valuation pursuant to clause 30. 2 [*substitute* '*clauses 4. 2. 1 and 4. 2. 2*' *when using IFC 98*] but that he has received no response.

You will be aware that your obligation was to make an interim valuation and you did so. However, where there is disagreement, a further obligation is for you to respond, providing him with a statement in the same amount of detail as his application identifying the areas of disagreement. The contractor complains that this was not done.

Perhaps you would let me know the details so that I can respond to the contractor. If you did not provide the statement, where there is a disagreement, that appears to be a breach of contract although I am unclear about the contractor's remedy for such breach.

Yours faithfully

函件 155

如果承包商要求就工地外的材料付款，且没有雇主清单，致承包商

此信函仅适用于 JCT98 合同或 IFC98 合同

Letter 155

To contractor, if he is seeking payment for off-site materials and there is no employer's list

This letter is only suitable for use with JCT 98 or IFC 98

尊敬的先生：

你方于［填入日期］要求就工地外的材料付款的信函收悉，深表谢意。依据合同第 30.3 条［使用 IFC98 合同时，替换为“第 4.2.1(c)条”］，只有雇主已提供给你方并附在合同清单列表中规定的预计用于工程的材料、货物和物品，其价格金额才能被写入证明书。

雇主没有提供这样的列表，因此我方无权证明你方的要求。

你忠诚的

Dear Sir

Thank you for your letter dated [*insert date*] seeking payment for off-site materials. Under the provisions of clause 30.3 [*substitute* '*4.2.1(c)*' *when using IFC 98*] the value of materials, goods or items prefabricated for inclusion in the Works can only be included in certificates if such materials are included in a list which the employer has supplied to you and annexed to the contract bills.

The employer has not produced such a list and, therefore, I have no power to certify as you request.

Yours faithfully

函件 156

如果承包商提交工程图纸，致承包商

此信函不适用于 WCD98 合同

Letter 156

To contractor, if he submits drawings

This letter is not suitable for use with WCD 98

尊敬的先生：

你方于［填入日期］的来函并附上编号为［填入编号］的工程图纸各两份收悉，深表谢意。

［加上：］

现寄还工程图纸各一份，我方无意见。

［或：］

我方的意见如下：

［然后：］

合同规定你方的责任是提供图纸，并综合所有图纸及其他所需的文件来进行施工。你方不应认为此信函可以减轻你方责任［视情况，加上："我方的意见同样如此"］。所有的图纸我方各留一份备案。

你忠诚的

Dear Sir

Thank you for your letter dated [*insert date*] with which you enclosed two copies of each of drawings numbers [*insert numbers*].

[*Either*:]

I return one copy of each drawing herewith without comment.

[*Or*:]

My comments are as follows: [*list comments*]

[*Then*:]

It is your responsibility under the contract to supply these drawings and to coordinate them and all other documents required to execute the Works. This letter must not be construed so as to relieve you of that responsibility [*insert*, *if appropriate*: '*and my comments are so restricted*']. I have retained one copy of the drawings for my records.

Yours faithfully

函件 157

如果承包商按补充条款 S2 提交规定的图纸，致承包商

此信函仅适用于 WCD98 合同

Letter 157

To contractor, if he submits drawings under supplementary provision S2

This letter is only suitable for use with WCD 98

尊敬的先生：

你方于［填入日期］的来信并附上编号为［填入编号］的图纸各两份收悉，深表谢意。

［加上：］

现寄还图纸各一份，我方无意见。

［或：］

我方的意见如下：

［然后：］

对工程作出计划是你方的责任，这一责任包括检查及综合所有需要的图纸来进行施工。你方不应认为此信函能减少你方的责任，也不能把此信函当作任何指示。我方的意见也不具有减轻你方责任的效力，请你方注意补充条款第 S2.2 条的说明。我方已经留有一份图纸备案。

你忠诚的

Dear Sir

Thank you for your letter dated [*insert date*] with which you enclosed two copies of each of drawings numbers [*insert numbers*].

[*Either*:]

I return one copy of each drawing herewith without comment.

[*Or*:]

My comments are as follows: [*list comments*]

[*Then*:]

It is your responsibility to design the Works. That responsibility includes the checking and co-ordination of all drawings required to execute the Works. This letter must not be construed so as to relieve you of that responsibility. Neither is it to be construed as an instruction of any kind. My comments are so restricted and your attention is drawn to the provisions of supplementary provision S2. 2 in that regard. I have retained one copy of the drawings for my records.

Yours faithfully

函件 158

如果需要信息，但没有信息发布日程表，致承包商

此信函仅适用于 JCT98 合同或 IFC98 合同

Letter 158

To contractor, if requesting information and there is no information release schedule

This letter is only suitable for use with JCT 98 or IFC 98

尊敬的先生：

你方于［填入日期］的要求［填入所需信息的简要细节］的来信收悉，深表谢意。

信息的提供需要遵循合同第 5.4.2 条［使用 IFC98 合同时，替换为“第 1.7.2 条”］。依据此条，我方必须在适当的时间向你方提供图纸及指示，以便你方能够按期竣工。但如果进度缓慢，在决定何时向你方提供信息为合理时，我方也有权将此因素考虑在内。鉴于此，我方预计于［填入日期］向你方提供所需信息。

你忠诚的

Dear Sir

Thank you for your letter dated [*insert date*] requesting [*insert brief details of the information required*].

Provision of information is governed by clause 5.4.2 [*substitute '1.7.2' when using IFC 98*]. Under this clause, I am obliged to provide you with drawings and instructions at such times as will enable you to complete the Works by the date for completion. But if progress is slow, I am entitled to take account of the slow progress when considering when it is reasonably necessary for you to receive the information. With that in mind, I anticipate providing the information you require by [*insert date*].

Yours faithfully

函件 159

如果要求发布日程表上的信息，致承包商

此信函仅适用于 JCT98 合同和 IFC98 合同

Letter 159

To contractor, if requesting information on the information release schedule

This letter is only suitable for use with JCT 98 and IFC 98

尊敬的先生：

你方于［填入日期］的要求［填入要求信息的简要细节］的来函收悉，深表谢意。信息发布日期已经在信息发布日程中标明。

［填入下列之一：］

随函附上所需信息。

［或：］

很抱歉，我方忽略了这些特定的图纸/细节/日程表［视情况取舍］，现随函附上。我方注意到此信息并没有造成你方工地工期的延误。

［或：］

距离信息发布日期还有［填入天数］天。不过，如果可行我方将尽力在此日期之前提供信息。

你忠诚的

Dear Sir

Thank you for your letter dated [*insert date*] requesting [*insert brief details of the information required*]. The release dates are given on the information release schedule.

[*Either*:]

The information is enclosed herewith.

[*Or*:]

I regret having overlooked these particular drawings/details/schedules [*delete as appropriate*] and I enclose them herewith. I note that lack of this information has not actually delayed you on site.

[*Or*:]

The release date is still some [*insert number*] days away. However, I will endeavour to provide the information before that date if practicable.

Yours faithfully

函件 160

如果信函意思不清，致承包商

Letter 160

To contractor, if letter not understood

尊敬的先生：

感谢你方于［填入日期］的来信。

我方阅读此信函几遍后，发现其中意思不清。如果我方就此回信，那么就只能是猜测你方的需要，这样只会把时间浪费在无用的事情上。欣然希望你方能重新措辞来函，以便我方能给予此信恰当的考虑。

你忠诚的

Dear Sir

Thank you for your letter of the [*insert date*].

After reading the contents several times, I am afraid that I find the meaning obscure. If I attempted to answer the letter as it stands, I should be merely guessing what you required of me. Rather than waste time on a fruitless exercise, I should be glad if you would rephrase the letter so that I can give it proper consideration.

Yours faithfully

函件 161

致承包商，暂时不对其问题给予详细答复

Letter 161

To contractor, pending detailed reply

尊敬的先生：

感谢你方于［填入日期］的来信。

鉴于信函内容的复杂性，我方不能立即详细答复你方来信的问题。我方会仔细考虑你方提到的事项，并及时致函你方。此信函不应该作为对你方信函的明确或暗示的同意。

你忠诚的

Dear Sir

Thank you for your letter of the [*insert date*].

In view of the complex nature of the contents, I am unable to reply in detail immediately. I will give careful consideration to the points you make and write to you again in due course. This letter is not to be taken as agreement, express or implied, to the whole or any part of your letter.

Yours faithfully

函件 162

致承包商，索要样品

此信函仅适用于 WCD98 合同

Letter 162

To contractor, requesting samples

This letter is only suitable for use with WCD 98

尊敬的先生：

我方注意到你方即将开工进行［描述所指的工程部分］。请你方注意，你方并未按《雇主要求》/《承包商建议书》［视情况取舍］第［填入参考段数］段，提供［填入材料］样品。

欣然希望你方能寄来样品。如果你方不提供样品而继续施工，就违反了合同条件第 8.6 条。

你忠诚的

Dear Sir

I note that you are about to commence work on [*describe the portion of work in question*] and I draw your attention to the fact that you have not yet complied with paragraph [*insert reference*] of the Employer's Requirements/Contractor's Proposals [*delete as appropriate*] which requires you to provide samples of [*insert material*].

I should be pleased to have these samples forthwith. If you proceed with the work before providing such samples, you will be in breach of clause 8.6 on the conditions of contract.

Yours faithfully

函件 163
关于样品事宜，致承包商

Letter 163
To contractor, regarding samples

尊敬的先生：

你方于［填入日期］寄来了［填入材料］的样品，并为说明问题作了相关标记［填入参考条数］。我方已经对这些样品进行了检验/测试［视情况取舍］。

这些样品符合工程量清单［或工程详细说明，或当使用WCD98合同时的《雇主要求》或《承包商建议书》］的［填入参考条款］的规定。我方对其工艺标准/材料质量［视情况取舍］没有意见。我方将留一份样品，其他样品将寄到工地由工程监督员保管。

你忠诚的

抄送：工程监督员

Dear Sir

I have examined/had tests carried out on [*delete as appropriate*] the samples of [*insert material*] which you delivered to this office on the [*insert date*] and which were marked [*insert reference number*] for identification purposes.

I have no comment to make on the standard of workmanship/quality of materials [*delete as appropriate*] as demonstrated by the samples in accordance with bills of quantities [*or specification, or Employer's Requirements or Contractor's Proposals when using WCD 98*] reference [*insert reference*]. One sample is being retained in this office and the remaining samples are being sent to site in the care of the clerk of works.

Yours faithfully

Copy: Clerk of works

函件 164
关于样品不合格，致承包商

Letter 164
To contractor, regarding failure of samples

尊敬的先生：

你方于［填入日期］寄来了［填入材料］的样品，并为说明问题作了相关标记［填入参考条数］。我方已经对这些样本进行了检验/测试［视情况取舍］。

样品不符合合同要求，不能令人满意。请参看［在图纸、工程量清单或工程详细说明，或当使用 WCD98 合同时的《雇主要求》或《承包商建议书》中填入参照条款］。请务必随后寄来质量合格的样品。

你忠诚的

抄送：工程监督员

Dear Sir

I have examined/had tests carried out on [*delete as appropriate*] the samples of [*insert material*] which you delivered to this office on the [*insert date*] and which were marked [*insert reference*] for identification purposes.

The samples are not satisfactory, because they are not in accordance with the contract. I refer you to [*insert appropriate reference in drawings, bills of quantities or specification, or Employer's Requirements or Contractor's Proposals when using WCD 98*]. Further samples of the proper quality must be submitted forthwith.

Yours faithfully

Copy: Clerk of works

函件 165

如果工地产品出现问题，致制造商

Letter 165

To manufacturer, if problems with product on site

尊敬的先生：

我方指定在以上项目中使用你方的［填入产品名称］。工程已经进行到［填入阶段］，我方注意到［填入出现的问题性质］。

［填入下列之一：］

现定于［填入日期、时间］在工地召集现场会议，请你方派代表参加。

［或：］

请派你方技术代表电话联系我方，由我方安排陪同到工地查看，以便商定解决的办法。

［或：］

我方打算/已经［视情况取舍］在大部分工程计划上使用此产品。除非你方于［填入日期］之前寄来解决问题的建议书，否则我方将修改我方的制定材料使用准则。

你忠诚的

Dear Sir

I specified your [*insert name of product*] on the above project. Work has reached [*insert stage*] and I am concerned to note that [*insert nature of problem*].

[*Add one of the following*:]

A site meeting is being held at [*insert time*] on [*insert date*]. Please arrange for your representative to be present.

[*Or*:]

Please arrange for your technical representative to telephone to arrange a visit to site in my company in order to suggest solutions to the difficulty.

[*Or*:]

I had intended to use/I have used [*delete as appropriate*] this product on a very large programme of work. Unless I have your proposals to solve the problem in my hands by [*insert date*] I intend to revise my specifying policy.

Yours faithfully

函件 166

关于缺陷工程，致工料测量师

此信函不适用于 WCD98 合同

Letter 166

To quantity surveyor, regarding defective work

This letter is not suitable for use with WCD 98

尊敬的先生：

我方在以上工地发现了下列缺陷工程：

［填入缺陷工程细节列表以便工料测量师能够找到，并包括上几个月直到工程缺陷被整改之前的各项记录］

以上工程在下次评估中将被删除。

你忠诚的

Dear Sir

The following defective work has been noted on the above site:

[*Insert list of defective work in sufficient detail to enable the quantity surveyor to identify it and include items from previous months until the defects have been corrected*]

The above work is to be omitted from your next valuation.

Yours faithfully

函件 167a

关于计划付款的金额，雇主致承包商的信函草稿

此信函不适用于 WCD98 合同

Letter 167a

Draft letter from employer to contractor, amount proposed to be paid

This letter is not suitable for use with WCD 98

尊敬的先生：

谨提及［填入日期］签发的金额为［填入金额］的证明书。请将此信函作为依据第 30.1.1.3 条/第 30.8.2 条［使用 IFC98 合同时，替换为“第 4.2.3(a)/4.3(b)/4.6.1.2 条”；使用 MW98 合同时，替换为“第 4.4.1/4.5.1.2 条”；使用 GC/Works/1（1998）合同时，替换为“第 50A(1)条”，视情况取舍］的通知[1]，我方计划向你方付款的金额为［填入金额］。这是关于上次评估后进行的施工和供应材料等的付款。此金额的计算方式如下：［填入计算标准，如果金额与证明书上一致，只需注明参照证明书中的计算方式］。

你忠诚的

［1　此通知必须于证明书发出后 5 天内发出。］

Dear Sir

I refer to certificate dated [*insert date*] in the sum of £ [*insert amount*]. Take this as notice[1] under clause 30.1.1.3/30.8.2 [*substitute* '*4.2.3(a)/4.3(b)/4.6.1.2*' *when using IFC 98 or* '*4.4.1/4.5.1.2*' *when using MW 98 or* '*50A(1)*' *when using GC/Works/1 (1998), deleting options as appropriate*] that the amount of payment I propose to make is £ [*insert amount*]. The payment relates to work carried out and materials, etc. supplied since the date of the last valuation. The amount is calculated as follows: [*insert the basis of calculation, but if the amount to be paid is the same as in the certificate it is enough to refer to the calculation of the certificate*].

Yours faithfully

[1 *This notice must be served no later than 5 days after the issue or a certificate.*]

函件 167b

关于计划付款的金额，雇主致承包商的信函草稿

此信函仅适用于 WCD98 合同

Letter 167b

Draft letter from employer to contractor, amount proposed to be paid

This letter is only suitable for use with WCD 98

尊敬的先生：

谨提及你方于［填入日期］发出的金额为［填入金额］的申请书。请把此信函作为依据合同第 30.3.3 条的通知[1]，我方计划付款金额为［填入金额］。此金额的计算方式如下：［填入计算标准，如果金额与申请书上一致，只需注明参照申请书中的计算方式］。

［如果有关于最终账单和最终报表的，则替换为以下内容］

最终报表于［填入日期］最终确定结欠金额。请把此信函作为依据合同第 30.6.1 条的通知[2]，我方计划付款金额为［填入金额］。此金额的计算方式如下：［填入计算标准，但如果金额是按照承包商的最终账单和最终报表计算，则只需说明参照内容］。

你忠诚的

［1 此通知必须于收到申请书后 5 天内发出。］

［2 此通知必须于最终报表确定结欠金额后 5 天内发出。］

Dear Sir

I refer to your application dated [*insert date*] in the sum of £ [*insert amount*]. Take this as notice[1] under clause 30. 3. 3 that the amount of payment I propose to make is £ [*insert amount*]. The amount is calculated as follows: [*insert the basis of calculation, but if the amount to be paid is the same as in the application it is enough to refer to the calculation of the application*].

[*If relating to the final account and final statement, put the following instead*:]

The final statement became conclusive as to the balance due on [*insert date*]. Take this as notice[2] under clause 30. 6. 1 that the amount of payment I propose to make is £ [*insert amount*]. The amount is calculated as follows: [*insert the basis of calculation, but if the amount to be paid is based on the contractor's final account and final statement, simply make reference to that*].

Yours faithfully

[1 *This notice must be served no later than 5 days after the receipt of an application for payment.*

2 *This notice must be served no later than 5 days after the final statement becomes conclusive about the balance due.*]

函件 168a

关于计划扣除的金额，雇主致承包商的信函草稿

此信函不适用于 WCD98 合同

Letter 168a

Draft letter from employer to contractor, amount proposed to be withheld

This letter is not suitable for use with WCD 98

尊敬的先生：

谨提及［填入日期］签发的金额为［填入金额］的证明书。请将此信函作为依据第 30.1.1.4 条/第 30.8.3 条［使用 IFC98 合同时，替换为“第 4.2.3(b)/4.3(c)/4.6.1.3 条”；使用 MW98 合同时，替换为“第 4.4.2/4.5.1.3 条”；使用 GC/Works/1 (1998) 合同时，替换为“第 50A(2)条”。视情况取舍］的通知[1]，我方计划从此金额中扣除［填入扣除金额］。扣除付款的依据以及按每一项依据扣除的金额如下：

［填入扣除的依据和按每一项依据扣除的金额。应较精确地计算］

你忠诚的

［1 此通知必须于最终付款日期 5 天前发出。］

Dear Sir

I refer to certificate dated [*insert date*] in the sum of £ [*insert amount*]. Take this as notice[1] under clause 30.1.1.4/30.8.3 [*substitute '4.2.3(b)/4.3(c)/4.6.1.3' when using IFC 98 or '4.4.2/4.5.1.3' when using MW 98 or '50A(2)' when using GC/Works/1 (1998) deleting options as appropriate*] that from that sum I propose to withhold £ [*insert amount to be withheld*]. The grounds for withholding and the amount withheld in respect of each ground are as follows:

[*Insert the grounds and the calculation of the amount alongside each ground. This should be done with some precision*]

Yours faithfully

[1 *This notice must be served no later than 5 days before the final date for payment.*]

函件 168b

关于计划扣除的金额，雇主致承包商的信函草稿

此信函仅适用于 WCD98 合同

Letter 168b

Draft letter from employer to contractor, amount proposed to be withheld

This letter is only suitable for use with WCD 98

尊敬的先生：

谨提及你方于［填入日期］发出的金额为［填入金额］的申请书。请把此信函作为依据第 30.3.4 条的通知[1]，我方计划从此金额中扣除［填入扣除金额］。扣除付款的依据以及按每一项依据扣除的金额如下：

［填入扣除的理由和按每一项理由扣除的金额。应较精确地计算］

［如果有关于最终账单和最终报表的，则替换为以下内容］

最终报表于［填入日期］最终确定结欠金额。请把此信函作为依据合同第 30.6.2 条的通知[2]，我方计划从结欠金额中扣除［填入扣除金额］。扣除付款的依据以及按每一项依据扣除的金额如下：

［填入扣除的依据和按每一项依据扣除的金额。应较精确地计算］。

你忠诚的

［1 此通知必须于最终付款日期 5 天前发出。］

［2 此通知必须于最终报表确定结欠金额并向承包商结清付款的最终日期 5 天前发出。］

Dear Sir

I refer to application dated [*insert date*] in the sum of £ [*insert amount*]. Take this as notice[1] under clause 30. 3. 4 that from that sum I propose to withhold £ [*insert amount to be withheld*]. The grounds for withholding and the amount withheld in respect of each ground are as follows:

[*Insert the grounds and the calculation of the amount alongside each ground. This should be done with some precision*]

[*If relating to the final account and final statement, put the following instead:*]

The final statement became conclusive as to the balance due on [*insert date*]. Take this as notice[2] under clause 30. 6. 2 that from such balance I propose to withhold £ [*insert amount to be withheld*]. The grounds for withholding and the amount withheld in respect of each ground are as follows:

[*Insert the grounds and the calculation of the amount alongside each ground. This should be done with some precision*]

Yours faithfully

[1 *This notice must be served no later than 5 days before the final date for payment.*

2 *This notice must be served no later than 5 days before the final date for payment of any balance in the final statement due to the contractor.*]

函件 169

如果缺陷工程被打开，致承包商

Letter 169

To contractor, if defective work opened up

尊敬的先生：

我方于［填入日期、时间］与［如果有的话，填入承包商代表及工程监督员的姓名］一起参加了［具体描述工程］的打开检验。

打开后发现工程与合同规定不符，按合同第 8.4 条［使用 IFC98 合同时，替换为“第 3.14 条”；使用 GC/Works/1（1998）合同时，替换为“第 40(2)(d)条”；使用 MW98 合同时，省略此项］，现随函附上一份指示要求你方进行清拆。依据第 8.3 条［使用 IFC98 合同时，替换为“第 3.12 条”；使用 GC/Works/1（1998）合同时，替换为“第 31(4)条和第 31(5)条”；使用 MW98 合同时，省略此项］，工程打开检验及整改的费用由你方支付。若你方认为工程已经按合同要求施工，你方应该在整改之前允许工程监督员/我方［视情况取舍］进行检查。

你忠诚的

抄送：工料测量师

工程监督员［如果已经委任］

Dear Sir

Together with [*insert name of contractor's representative and clerk of works, if any*], I attended the opening up of [*specify precisely*] at [*insert time*] on [*insert date*].

The work was found to be not in accordance with the contract and an instruction is enclosed under clause 8.4 [*substitute '3.14' when using IFC 98, '40(2)(d)' when using GC/Works/1 (1998) or omit the phrase when using MW 98*] requiring removal. The cost of opening up and making good is to be at your expense in accordance with clause 8.3 [*substitute '3.12' when using IFC 98, '31(4) and 31(5)' when using GC/Works/1 (1998) or omit the phrase when using MW 98*]. When you consider that the work has been executed in accordance with the contract, the clerk of works/I [*delete as appropriate*] must be allowed to inspect before making good takes place.

Yours faithfully

Copies: Quantity surveyor
Clerk of works [*if appointed*]

函件 170a

发现工程不合格后，致承包商

此信函仅适用于 JCT98 和 WCD98 合同

Letter 170a

To contractor, after failure of work

This letter is only suitable for use with JCT 98 and WCD 98

尊敬的先生：

今天查看工地时，我方注意到［具体指明工程或材料］与合同要求不符。

依据合同条件第 8.4.4 条［使用 WCD98 合同时，替换为“第 8.4.3 条”］，并遵循此条件的惯例，现附上打开工程进行检验/测试［视情况取舍］的指示。此检查是为了证实工程是否有类似于不符合合同要求的可能性/程度［视情况取舍］。

你忠诚的

Dear Sir

While visiting site today, I noticed that [*specify work or materials*] were not in accordance with the contract.

In accordance with clause 8.4.4 [*substitute '8.4.3' when using WCD 98*] of the conditions of contract and having had due regard to the code of practice appended to such conditions, I enclose my instruction for opening up for inspection/testing [*delete as appropriate*] which is reasonable in all the circumstances to establish to my reasonable satisfaction the likelihood/extent [*delete as appropriate*] of any similar non-compliance.

Yours faithfully

函件 170b

发现工程不合格后，致承包商

此信函仅适用于 IFC98 合同

Letter 170b

To contractor, after failure of work

This letter is only suitable for use with IFC 98

尊敬的先生：

今天查看工地时，我方注意到［具体指明工程或材料］与合同要求不符。

按合同条件第 3. 13. 1 条，现要求你方于收到此信函 7 天内来函告知你方将立刻采取的任何行动，且此项费用不由雇主支付，以便证实已经施工的工程/已经供应的材料或货物［视情况取舍］并不存在类似的不合格情况。

你忠诚的

Dear Sir

While visiting site today, I noted that [*specify work or materials*] were not in accordance with the contract.

In accordance with clause 3. 13. 1 of the conditions of contract, I require you to state in writing within 7 days of the date of this letter what action you will immediately take at no cost to the employer to establish that there is no similar failure in work already executed/materials or goods already supplied [*delete as appropriate*].

Yours faithfully

函件 171

致委托人，确认需要额外支付费用的指示

Letter 171

To client, confirming instruction which entails extra cost

尊敬的先生：

谨提及今晨与你方的电话谈话。你方指示我方安排进行如下工程［具体说明该工程：］

现确认：

1. 此项工程预计造价为［填入金额］。实际费用将加到合同总额之中。
2. 你方批准此费用，并希望我方指示承包商尽快着手进行此工程。

你忠诚的

抄送：工料测量师

Dear Sir

I refer to our telephone conversation this morning when you instructed me that the following work was to be carried out: [*specify work*]

I confirm that:

1. The estimated cost of the work will be [*insert amount*]. The actual amount will be added to the contract sum.
2. You approve this cost and wish me to instruct the contractor to put the work in hand as soon as possible.

Yours faithfully

Copy: Quantity surveyor

函件 172

如有必要对已经批准的设计更换材料，致委托人

Letter 172

To client, if material change to approved design necessary

尊敬的先生：

你方已经于［填入日期］批准了我方对此项目的设计。从那时起［描述发生的重大事情］。我方认为有必要［描述设计的更改］。此更改是对已经批准的设计更换材料。此信函是依据雇佣条款第 2.7 条［使用 SW/99 合同时，替换为“第 6 条”］发出的，以便向你方确认，在我们于［填入日期］的面议时，你方已同意此项变更。

你忠诚的

Dear Sir

You approved my design for this project on [*insert date*]. Since that time [*insert description of the significant event*]. It is my view that it is necessary to [*describe change in design*] which is a material change from the approved design. This letter is issued in accordance with clause 2.7 [*substitute '6' when using SW/99*] of the terms of engagement to confirm that, when we met on [*insert date*], you consented to the change.

Yours faithfully

函件 173

致承包商，要求对方遵循指示施工

专递/挂号邮件

Letter 173

To contractor, requiring compliance with instruction

Special/recorded delivery

尊敬的先生：

依据合同条件第 4.2.1 条［使用 IFC98 合同时，替换为“第 3.5.1 条”；使用 MW98 合同时，替换为“第 3.5 条”；使用 GC/Works/1（1998）合同时，替换为“第 53 条”］，此信函作为正式通知要求你方遵循我方于［填入日期］发出的［填入编号］指示。现再次随函附上此指示副本。

如果收到此信函 7 天［使用 GC/Works/1（1998）合同时，替换为任何合适的期限］内，你方仍未遵循其施工，雇主可能会雇佣其他承包商进行此指示要求的工作并向对方付费。与之相关的任何费用将从合同规定支付你方或将支付你方的费用中扣除，或作为你方的债务得到偿付。

你忠诚的

抄送：雇主

工料测量师

Dear Sir

Take this as notice under clause 4. 1. 2 [*substitute* '*3. 5. 1*' *when using IFC 98*, '*3. 5*' *when using MW 98 or* '*53*' *when using GC/Works/1* (*1998*)] of the conditions of contract that I require you to comply with my instruction number [*insert number*] dated [*insert date*], a further copy of which is enclosed.

If within 7 [*substitute any reasonable period when using GC/Works/1* (*1998*)] days of receipt of this notice you have not complied, the employer may employ and pay others to execute any work necessary to give effect to the instruction. All costs incurred in connection with such employment will be deducted from money due or to become due to you under the contract or will be recovered from you as a debt.

Yours faithfully

Copies: Employer
Quantity surveyor

函件 174

如果承包商没有服从通知，致承包商

专递/挂号邮件

Letter 174

To contractor, if he fails to comply with notice

Special/recorded delivery

尊敬的先生：

谨提及我方于［填入日期］发出的通知，依据合同第 4.1.2 条［使用 IFC98 合同时，替换为“第 3.5.1 条”；使用 MW98 合同时，替换为“第 3.5 条”使用 GC/Works/1（1998）合同时，替换为“第 53 条”］，要求你方遵循［填入日期］发出的［填入编号］的指示施工。

今晨对工地进行检查时，我方注意到你方并未遵循我方的指示施工，雇主将立即着手雇佣其他承包商来进行此项工作。与之相关的任何费用将从合同规定支付你方或将支付你方的费用中扣除，或作为你方的债务得到偿付。

你忠诚的

抄送：雇主
工料测量师

Dear Sir

I refer to the notice issued to you on the [*insert date*] in accordance with clause 4.1.2 [*substitute '3.5.1' when using IFC 98, '3.5' when using MW 98 or '53' when using GC/Works/1 (1998)*] requiring compliance with my instruction number [*insert number*] dated [*insert date*].

During a site inspection this morning, I noted that you have not complied with my instruction. The employer is taking immediate steps to employ others to carry out the work. All costs in connection with such employment will be deducted from money due or to become due to you under the contract or will be recovered from you as a debt.

Yours faithfully

Copies: Employer
Quantity surveyor

函件 175

如果没有依据而意图对工程延期，致承包商

Letter 175

To contractor, if no grounds for extension of time

尊敬的先生：

你方于［填入日期］的来信收悉，深表谢意。来信中你方坚持，按你方列出的若干依据，你方有权要求工程延期。

按照你方提交的文件和我方对此项目的了解，我方认为你方没有理由延期竣工。如果你方能按合同条款规定并以适当形式提交延期依据，我方将欣然予以考虑。

你忠诚的

Dear Sir

Thank you for your letter of the [*insert date*] in which you asserted that you were entitled to an extension of time for the reasons set out.

On the basis of the documents you have presented to me and of my own knowledge of this project, I can see no grounds for any extension of time. I shall be pleased to consider any further submissions if they are presented in the proper form and in accordance with the terms of the contract.

Yours faithfully

函件 176
如果不应当延期，致承包商

Letter 176
To contractor, if no extension of time due

尊敬的先生：

我方仔细审查了你方的延期通知及相关细节。我方认为你方不应该要求延期/再次延期［视情况取舍］。

你忠诚的

Dear Sir

I have carefully examined your notice of delay and accompanying particulars and it is my opinion that you are not due an extension of time/further extension of time [*delete as appropriate*].

Yours faithfully

函件 177

由于竣工日期前时间不足，需要分两次发出延期通知，致承包商

Letter 177

To contractor, if issuing extension in two parts because of lack of time before completion date

尊敬的先生：

我方已经于［填入日期］收到你方的延期通知和相关依据，打算给予适当的延期批准。

鉴于合同规定的竣工日期临近，且考虑此事的各方面因素需要花费一定的时间，依据我方对你方证据初步的审查，现附上一份临时延期证明。

一旦我方对此事调查结束后，你方可能会立刻得到再次延期批准［视情况替换为“在合同规定的期限内”］。

你忠诚的

Dear Sir

I have received your delay notices and supporting information on the [*insert date*] and I have now to consider an appropriate extension of time.

In view of the proximity of the contractual completion date and the length of time it will take to consider all aspects of the matter, I enclose an interim extension based upon my initial view of the evidence.

Any further extension of time to which you may be entitled will be given as soon as my investigations are concluded [*substitute 'within the period stipulated in the contract' if appropriate*].

Yours faithfully

函件 178

致承包商，给予延期批准

Letter 178

To contractor, giving extension of time

尊敬的先生：

谨提及你方于［填入日期］发出的延期通知［视情况，加上：］及［填入日期］信中提供的进一步相关信息。

依据合同第25.3条［使用IFC98合同时，替换为“第2.3条”；使用MW98合同时，替换为“第2.2条”；使用GC/Work/1（1998）合同时，替换为“第36（1）条”］，现给予你方［填入期限］的延期批准。修改后的竣工日期为［填入日期］。

［使用JCT98合同或IFC98合同时，加上：］

需要考虑的相关事项有：［列出］

［使用JCT98合同时，也要加上：］

我方注意到以下要求删除工程的变更指示。

［列出指示和减少延期的程度，如：

建筑师指示	延期中删除
No. 24, 03. 09. 95	14天］

［使用GC/Works/1（1998）合同时，加上：］

此决定为临时/最终［视情况取舍］决定。

你忠诚的

Dear Sir

I refer to your notice of delay dated [*insert date and if appropriate, add:*] and the further information provided in your letter dated [*insert date*].

In accordance with clause 25.3 [*substitute '2.3' when using IFC 98, '2.2' when using MW 98 or '36(1)' when using GC/Works/1 (1998)*] I hereby give you an extension of time of [*insert period*]. The revised date for completion is now [*insert date*].

[*When using JCT 98 or IFC 98, add:*]

The relevant events taken into account are: [*list*]

[*When using JCT 98, add also:*]

I have had regard to the following instructions requiring as a variation the omission of work:

[*List the instructions and the extent of any reduction in extension of time, e.g:*
Architect's Instruction *Omission from extension*
No. 24, 03.09.95 *14 days*]

[*When using GC/Works/1 (1998), add:*]

This is an interim/final [*delete as appropriate*] decision.

Yours faithfully

函件 179

致委托人，随函附上工程延期报告

Letter 179

To client, enclosing a report on extension of time

尊敬的先生：

承包商通知我方工程可能会延期，并估计了竣工日期受到的影响。

在决定是否批准承包商要求延期的权限时，我方必须严格遵循合同条款。

现附上一份关于此事的简要报告，其中说明了主要的涉及因素及我方对其的每项决定。此报告供你方参考，如有意义不清之处，我方将欣然为你方解释。

你忠诚的

Dear Sir

The contractor has notified me of likely delay and he has estimated the effect on the completion date.

In making my decision on any entitlement to extension of the contract period, I must act strictly in accordance with the terms of the contract.

I enclose a copy of my brief report on the matter in which I have set out the main factors and my decision in each instance. This report is for your information, but I will be happy to explain anything which may seem unclear.

Yours faithfully

函件 180

如果工程未能按期竣工，可能需要再次延期，致委托人

此信函不适用于 MW98 合同或 GC/Works/1（1998）合同

Letter 180

To client, if Works not complete by due date and further extensions may be due

This letter is not suitable for use with MW 98 or GC/Works/1 (1998)

尊敬的先生：

本合同的竣工时间为［填入日期］。现随函附上合同第 24.1 条［使用 IFC98 合同时，替换为“第 2.6 条”］规定的未能竣工证明书。

如果你方愿意，你方可以按附件中规定的利率扣除从合同规定竣工时间到实际竣工时间的违约金。或者你方可以把此作为承包商的债务而获得偿付。如果你方决定了上述方案之一，就必须首先给承包商发出书面的要求，说明扣除金额涉及的期限以及违约金的计算方法。然后你方必须寄去恰当的扣除通知。

在决定是否扣除预定违约金的同时，请你方记住，我方要进行关于工程延期的审查。等到工程实际竣工之后再开始审查是不合适的。工程有可能再次延期。如果我方复查后，发现应当再次给予延期，我方会确定新的竣工时间，你方需要将到新的竣工日期为止扣除的预定违约金归还承包商。

你忠诚的

Dear Sir

The due date for completion of this contract was [*insert date*] and I enclose my certificate of non-completion as required by clause 24. 1 [*substitute* '*2. 6*' *when using IFC 98*] of the contract.

You may, if you wish, deduct liquidated damages at the rate stated in the appendix for the period between the date the contract should have been completed and the date of practical completion. Alternatively, you may recover the damages as a debt. If you decide upon either of these routes, you must first give the contractor a written requirement, setting out the period to which the deduction relates and the way in which the damages are calculated. You must follow this with the appropriate withholding notice.

Please bear in mind, when deciding whether or not to deduct liquidated damages, that I have yet to carry out my review of extensions of time. It is not appropriate to begin until practical completion has occurred. Further extensions may be due. If, as a consequence of my review, I fix a later date for completion, you would be liable to repay any liquidated damages deducted for the period up to the later completion date.

Yours faithfully

函件 181a

致承包商，在竣工日期或实际竣工后进行工程延期的复审

此信函仅适用于 JCT98 合同或 WCD98 合同

Letter 181a

To contractor, reviewing extensions of time after completion date or practical completion

This letter is only suitable for use with JCT 98 and WCD 98

尊敬的先生：

我方已经按合同条件第 25. 3. 3 条对工程进度进行了考查。

[或：]

我方现确认合同的竣工日期为 [填入日期]。

[或：]

我方考虑到 [填入相关事宜]，现给予 [填入期限] 的工程延期批准。新的竣工时间为 [填入日期]。

[或：]

我方注意到上次延期后发出了以下要求删除工程的 [填入编号] 指示，新的竣工日期为 [填入合理的稍早的日期，但不能早于附件中的规定日期]。

你忠诚的

抄送：雇主

工料测量师

Dear Sir

I have now completed my consideration of progress in accordance with clause 25. 3. 3 of the conditions of contract and

[*Either*:]

I confirm the date of completion of the contract as being [*insert date*].

[*Or*:]

I hereby give an extension of time of [*insert period*] which takes into account [*insert relevant events*]. The new date for completion is [*insert date*].

[*Or*:]

After having regard to the following instructions [*insert instruction numbers*] requiring omissions and issued after the last occasion on which I made an extension of time, the new date for completion is now [*insert such earlier date as is reasonable but not earlier than the date in the appendix*].

Yours faithfully

Copies: Employer
Quantity surveyor

函件 181b

致承包商，在竣工日期或实际竣工后进行工程延期的复审

此信函仅适用于IFC98合同

Letter 181b

To contractor，reviewing extensions of time after completion date or practical completion

This letter is only suitable for use with IFC 98

尊敬的先生：

我方已经按合同条件第2.3条对工程进度进行了考查。

［或：］

我方现确认合同的竣工日期为［填入日期］。

［或：］

我方考虑到［填入相关事宜］，现给予［填入期限］的工程延期批准。新的竣工时间为［填入日期］。

你忠诚的

抄送：雇主
工料测量师

Dear Sir

I have now completed my consideration of progress in accordance with clause 2. 3 of the conditions of contract and

[*Either*:]

I confirm the date of completion of this contract is [*insert date*].

[*Or*:]

I hereby give an extension of time of [*insert period*] which takes into account [*insert reasons for extension*]. The new date for completion is [*insert date*].

Yours faithfully

Copies: Employer
Quantity surveyor

函件 182a

致委托人，随函附上工程未能竣工证明书

此信函不适用于 MW98 合同或 GC/Works/1（1998）合同

Letter 182a

To client, enclosing certificate of non-completion

This letter is not suitable for use with MW 98 or GC/Works/1 (1998)

尊敬的先生：

随函附上按合同条件第 24.1 条［使用 IFC98 合同时，替换为“第 2.6 条”］规定发出的证明书［使用 WCD98 合同时，替换为“通知”］。

你方可以现在按合同附件中规定的利率索取预定违约金。最简单的办法就是从我方发出的财务证明［使用 WCD98 合同时，替换为“承包商付款申请书”］规定应付给承包商的金额中扣除相应部分。如果你方决定采取这种方式，就必须在扣除之前致函承包商，以便他方能够确定你方将扣除违约金，并明确扣除金额的计算方法。我方已经随函附上了此信函的草稿。每次你方打算扣除预定违约金时，都应该以类似方式通知承包商，另外你方必须寄去一份有效的扣除通知。

你忠诚的

Dear Sir

I enclose my certificate [*substitute 'notice' when using WCD 98*] in accordance with clause 24.1 [*substitute '2.6' when using IFC 98*] of the conditions of contract.

You may now take steps to recover liquidated damages at the rate stated in the appendix to the contract. The easiest way to do this is to deduct the appropriate sum from amounts otherwise due under my financial certificates [*substitute 'the contractor's applications for payment' when using WCD 98*]. If you decide to follow this course of action, I enclose a draft letter which you must write to the contractor before making the deduction so that he is in no doubt that you intend to make the deduction and the way in which the amount of the deduction is calculated. You should inform him in a similar manner before each occasion on which you intend to deduct liquidated damages. In addition, you must send an effective withholding notice.

Yours faithfully

函件 182b

如果工程没有竣工，致委托人

此信函仅适用于 MW98 合同或 GC/Works/1（1998）合同

Letter 182b

To client, if Works not complete

This letter is only suitable for use with MW 98 or GC/Works/1（1998）

尊敬的先生：

合同条件规定的竣工日期/第 2.2 条规定的竣工日期/第 36 条规定的竣工日期［视情况取舍］是［填入日期］。谨以此信函告知你方工程没有竣工。

你方可以现在按合同第 2.3 条规定的［使用 GC/Works/1（1998）合同时，替换为“专用条款中说明”］利率索取预定违约金。最简单的办法就是从我方发出的财务证明规定应付给承包商的金额中扣除相应部分。如果你方决定采取这种方式，就必须在扣除之前致函承包商，以便他方能够确定你方将扣除违约金，并明确扣除金额的计算方法。我方已经随函附上了此信函的草稿。每次你方打算扣除预定违约金时，都应该以类似方式通知承包商，另外你方必须寄去一份有效的扣除通知。

你忠诚的

Dear Sir

The date for completion/completion date fixed under clause 2. 2/completion date fixed under clause 36［*delete as appropriate*］of the conditions of contract is［*insert date*］. This letter is to inform you that the Works are not complete.

You may now take steps to recover liquidated damages at the rate stated in clause 2. 3 of the contract［*substitute* ‘*stated in the abstract of particulars*’ *when using GC/Works/1（1998）*］. The easiest way to do this is to deduct the appropriate sum from amounts otherwise due under my financial certificates. If you decide to follow this course of action, I enclose a draft letter which you should write to the contractor before making the deduction so that he is in no doubt that you intend to make the deduction and the way in which the amount of the deduction is calculated. You should also send an effective withholding notice.

Yours faithfully

函件 183

扣除违约金前，雇主致承包商的信函草稿

专递/挂号邮件

Letter 183

Draft letter from employer to contractor, before deducting liquidated damages

Special/recorded delivery

尊敬的先生：

按照合同第 24.2.1 条［使用 IFC98 合同时，替换为“第 2.7 条”；使用 MW98 合同时，替换为“第 2.3 条”；使用 GC/Works/1（1998）合同时，替换为“第 55 条”］，我方要求你方支付预定违约金，或者我方将从下次财务证明中［使用 WCD98 合同时，替换为“付款书”］扣除此金额。

你忠诚的

Dear Sir

In accordance with clause 24.2.1 [*substitute* '2.7' *when using IFC 98*, '2.3' *when using MW98 or* '55' *when using GC/Works/1 (1998)*] I may require you to pay or I may deduct liquidated damages from the next financial certificate [*substitute* '*payment*' *when using WCD 98*].

Yours faithfully

函件 184

关于扣除预定违约金的建议，致委托人

Letter 184

To client, advising on the deduction of liquidated damages

尊敬的先生：

你方有权按适当的利率扣除预定违约金。合同明确规定是否扣除违约金完全取决于你方的意愿。尽管我方愿意给予意见，但可能有一些我方不了解的因素会影响你方的决定。此违约金目前的计算方法如下：

合同规定的竣工日期：［填入日期或延期后的日期］

预定违约金计算的日期：［填入日期，并视情况说明此日期是否为实际竣工或竣工日期］。

=［填入数字］周@［填入金额］每［填入单位，如天数或周数］=［填入总金额］

扣除通知中也应该包括此计算方法。

你忠诚的

Dear Sir

You are entitled to deduct liquidated damages at the appropriate rate. The contract stipulates that it is a matter for your discretion alone. Although I am always ready to give you advice, there may be considerations, unknown to me, which will influence your decision. Calculation of such damages is currently as follows:

Contract date for completion: [*insert date or extended date*]

Date at which liquidated damages calculated: [*insert date and state if it is also the date of practical completion or completion as appropriate*]

= [*insert number*] weeks @ [*insert amount*] per [*insert unit, e.g. days or weeks*] = [*insert total sum*].

This calculation should be included in the withholding notice.

Yours faithfully

函件 185

如果扣除预定违约金不公平，致委托人

Letter 185

To client, if it would be unfair to deduct liquidated damages

尊敬的先生：

此合同的竣工日期为［填入日期］，但承包商尚未竣工。现随函附上合同要求的未竣工证明［使用 WCD98 合同时，替换为“通知”］［视情况取舍］。

你方有权按适当的利率扣除预定违约金。合同明确规定是否扣除违约金完全取决于你方的意愿。

可能有一些我方不了解的因素会影响你方的决定，但请你方注意以下事项：

［代表承包商填入情有可原的信息］

你方做决定时，可以考虑或不考虑这些信息。

你忠诚的

Dear Sir

The due date for completion of this contract was [*insert date*] and the contractor has not yet completed the Works. I enclose my certificate [*substitute* '*notice*' *when using WCD 98*] of non-completion as required by the contract [*delete as appropriate*].

You are entitled to deduct liquidated damages at the appropriate rate. The decision to deduct is a matter for your discretion alone.

There may be considerations, unknown to me, which will influence your decision. May I, however, draw your attention to the following:

[*Insert any mitigating information on behalf of the contractor*]

You may or may not wish to take this information into account when making your decision.

Yours faithfully

函件 186

致承包商，同意给予指定分包商工程延期批准

此信函仅适用于 JCT98 合同

Letter 186

To contractor, giving consent to extension of time for nominated sub-contract

This letter is only suitable for use with JCT 98

尊敬的先生：

我方于［填入日期］收到你方依据合同第 35. 14 条发出的关于由［填入分包商姓名］进行指定分包合同施工的通知、细节说明和估价单。

依据合同第 35. 14 条和分包合同第 2. 3 条，我方同意你方建议的工程延期，这样分包工程的竣工期限即修改为［填入期限］。按分包合同第 2. 3. 3 条，我方同意将以下事项列入考虑范围：［列出事项］。

［视情况，加上：］

在我方同意延期的同时，也注意到以下的要求删除工程的变更指示：［列出指示及减少延期期限的程度］。

你忠诚的

Dear Sir

I refer to your notice, particulars and estimate in accordance with clause 35. 14 of the contract which I received on the [*insert date*] in respect of the nominated sub-contract Works being carried out by [*insert name*].

In accordance with the provisions of clause 35. 14 of the contract and clause 2. 3 of the sub-contract I hereby give my consent to the extension of time you propose which gives a revised period for completion of the sub-contract Works of [*insert period*]. I agree that the matters taken into account in accordance with clause 2. 3. 3 of the sub-contract are as follows: [*list matters*].

[*If appropriate, add*:]

I further agree that, in giving my consent, I have had regard to the following instructions requiring as variations the omission of work: [*list the instructions and the extent by which they have reduced the period of extension*].

Yours faithfully

函件 187

致承包商，不同意指定分包商的工程延期

此信函仅适用于 JCT98 合同

Letter 187

To contractor, withholding consent to extension of time for nominated sub-contract

This letter is only suitable for use with JCT 98

尊敬的先生：

我方于［填入日期］收到你方依据合同第 35. 14 条发出的关于由［填入分包商姓名］进行指定分包合同施工的通知、细节说明和估价单。

我方不能同意你方建议的工程延期，因为我方认为该延期期限并不正确。

你忠诚的

Dear Sir

I refer to your notice, particulars and estimate submitted under the provisions of clause 35. 14 of the contract which I received on the [*insert date*] in respect of the nominated sub-contract Works carried out by [*insert name*].

I cannot give my consent to your proposed award of extension of time, because I consider the period is not correct.

Yours faithfully

函件 188

致承包商，不签发指定分包商未能竣工的证明书

此信函仅适用于 JCT98 合同

Letter 188

To contractor, withholding certificate of nominated sub-contractor's failure to complete

This letter is only suitable for use with JCT 98

尊敬的先生：

谨提及你方于［填入日期］按合同第 35. 15. 1 条提交的通知。

我方拒绝签发关于指定分包商［填入姓名］未能在分包合同规定的期限内/我方书面同意的延期时间内［视情况取舍］完成分包工程的证明书。因为：

［填入：］

我方对指定分包商［填入姓名］竟未能按时竣工表示遗憾。

［或：］

我方对你方使用合同第 35. 14 条表示不赞成。

你忠诚的

Dear Sir

I refer to your notice dated [*insert date*] and submitted under clause 35. 15. 1 of the contract.

I decline to issue my certificate that [*insert name of nominated sub-contractor*] has failed to complete the sub-contract Works within the period specified in the sub-contract/the extended time granted with my written consent [*delete as appropriate*], because

[*Insert either*:]

I am not satisfied that [*insert name of nominated sub-contractor*] has so failed.

[*Or*:]

I am not satisfied that clause 35. 14 has been properly applied.

Yours faithfully

函件 189

致承包商，证实指定分包商未能竣工

此信函仅适用于 JCT98 合同

Letter 189

To contractor, certifying nominated sub-contractor's failure to complete

This letter is only suitable for use with JCT 98

尊敬的先生：

谨提及你方于［填入日期］按合同第 35. 15. 1 条提交的通知。

我方证实［填入地址］的［填入指定分包商姓名］未能在分包合同规定的期限内/我方书面同意的由承包商授权的延期时间［填入期限］内［视情况取舍］完成分包工程

你忠诚的

抄送：指定分包商

Dear Sir

I refer to your notice dated [*insert date*] and submitted under clause 35. 15. 1 of the contract.

I certify that [*insert name of nominated sub-contractor*] of [*insert address*] has failed to complete the sub-contract Works within the period specified in the subcontract/the extended time granted by the contractor with my written consent [*delete as appropriate*] namely [*insert period*].

Yours faithfully

Copy: Nominated sub-contractor

函件 190

关于依据普通法的索赔，致委托人

此信函不适用于 MW98 合同

Letter 190

To client, regarding common law claims

This letter is not suitable for use with MW 98

尊敬的先生：

现随函附上主承包商于［填入日期］向我方发出的关于补充付款的申请书。

合同中仅提供了我方处理特定类别的财务索赔的条款。此申请书不属于/一部分不属于［视情况取舍］这些条款权限，因此我方无权按合同确定其有效性或决定是否付款。

我方建议你方目前不要简单地拒绝赔款，因为承包商可能会决定通过合同争议解决程序来处理此事。如果你方能授权给我方，我方乐意对申请书的合理性进行审查，并就付款事宜给你方建议。或者，你方可能希望获得精通建筑法律的专家的建议。如果是这样，我方将很乐意配合你方提供任何需要的补充信息。

你忠诚的

Dear Sir

I enclose an application for additional payment dated [*insert date*] sent to me by the main contractor.

The contract makes provisions for me to deal with financial claims of specific kinds. This application does not/parts of this application do not [*delete as appropriate*] fall within these provisions and I have no authority under the contract to ascertain either validity or payment.

I advise that you should not simply reject the claim at this stage, because the contractor may decide to pursue the matter through the contract dispute resolution process. If you give me the requisite authority, I am prepared to examine the application on its merits and advise you regarding payment. Alternatively, you may wish to obtain specialised advice from a person versed in construction law. If so, I should be happy to accompany you to provide any additional information required.

Yours faithfully

函件 191

关于损失和（或）费用申请，致委托人

此信函仅适用于 MW98 合同

Letter 191

To client regarding loss and/or expense applications

This letter is only suitable for use with MW 98

尊敬的先生：

谨随函附上承包商提交的［填入日期］发出的损失和（或）费用申请书。尽管合同第 3.6 条包括一个内容有限的条款，可以作为我方在评估指示中考虑损失和（或）费用的依据，但合同条款并未允许我方确认当前提交的申请书的有效性或决定是否付款。

请不要立即拒绝赔款，因为承包商可能会决定通过合同争议解决程序来处理此事。这样做会花费较高的费用。如果你方想就此事与我方商讨，请电话联系我方，以安排合适的会议时间。也可能会极为简单地与承包商解决此事。会议上，我们可以讨论是否有必要在采取法律程序之前寻求专家的意见。

你忠诚的

Dear Sir

I enclose an application for loss and/or expense dated [*insert date*] which has been submitted by the main contractor. Although there is a limited provision in clause 3.6 for me to include direct loss and/or expense in the valuation of instructions, the contract provisions do not allow me to ascertain validity or payment for submissions such as the enclosed.

Do not reject the claim out of hand, because the contractor may decide to pursue his claim through the contract dispute resolution procedures, which could be expensive. If you wish to discuss the claim with me, please telephone to arrange a suitable time for a meeting. It may be possible to settle the matter quite easily with the contractor. At our meeting, we could discuss whether it is necessary to obtain specialised advice before proceeding.

Yours faithfully

函件 192

关于通融索赔，致委托人

Letter 192

To client, regarding ex gratia claims

尊敬的先生：

现随函附上主承包商寄来的于［填入日期］向我方发出的付款申请书。

这完全是寻求通融索赔。也就是说，此要求并没有法律依据。承包商仅仅是要求你方考虑付款，因为他方目前财务困难，如果你方需要具体信息以便做出决定，请通知我方。请记住，即使你方确定承包商财务困难，你方也并无义务向他方付款。

你忠诚的

Dear Sir

I enclose an application for payment dated [*insert date*] sent to me by the main contractor.

The claim appears to be purely ex gratia. In other words, it has no legal basis. The contractor is simply asking you to consider making a payment, because he has suffered hardship. Let me know if you require any particular information in order to come to a decision. Remember that, even if you decide that the contractor has indeed suffered hardship, you are under no obligation to pay anything.

Yours faithfully

函件 193

如果承包商提交通融索赔申请或按普通法索赔申请，致承包商

Letter 193

To contractor, if ex gratia or common law application submitted

尊敬的先生：

我方已经收到你方于［填入日期］发出的付款申请，并已将之转交给雇主，以便他方作决定。因为根据合同条款，这不属于我方的职权范围。

你忠诚的

抄送：雇主

工料测量师

Dear Sir

I have received your application for payment dated [*insert date*] which has been passed to the employer for his decision, because it does not fall within the limits of my authority under the terms of the contract.

Yours faithfully

Copies: Employer

Quantity surveyor

函件 194

致承包商，拒绝承包商由于未遵循合同造成损失而发出的付款申请

此信函不适用于 MW98 合同

Letter 194

To contractor, rejecting application due to failure to comply with contract

This letter is not suitable for use with MW 98

尊敬的先生：

我方于［填入日期］［填入时间］收到你方发出的损失和（或）费用［使用 GC/Works/1（1998）合同时，替换为“费用”］申请书。

我方不能考虑你方的申请，因为你方没有遵循合同第 26 条［使用 IFC 98 合同时，替换为“第 4.11 条”；使用 GC/Works/1（1998）合同时，替换为“第 46 条”］的要求。而且，你方未履行职责［简单明了地陈述承包商的失职之处］。

你忠诚的

抄送：工料测量师

Dear Sir

I refer to your application for loss and/or expense [*substitute* '*expense*' *when using GC/Works/1 (1998)*] received on the [*insert date*].

I cannot begin to consider your claim, because you have failed to comply with the requirements of clause 26 [*substitute* '*4.11*' *when using IFC 98 or* '*46*' *when using GC/Works/1 (1998)*]. In particular, you have failed in your duty to [*specify, clearly and concisely, how the contractor has failed*].

Yours faithfully

Copy: Quantity surveyor

函件 195

如果申请书措辞不清，致承包商

此信函不适用于 MW98 合同

Letter 195

To contractor, if application badly presented

This letter is not suitable for use with MW 98

尊敬的先生：

谨提及你方于［填入日期］发出的损失和（或）费用［使用 GC/Works/1 (1998) 合同时，替换为“费用”］申请书。

尽管从表面上看，你方满足了合同所有的程序要求，我方可以开始考虑你方的申请，但你方的申请书措辞不清。我方打算按现状审查申请书，但我方必须告诫你方，如果我方对其有效性不满意，我方有责任否决此申请书或其中的部分内容。

请考虑是否重新提交更明晰易懂的申请书。我方将推迟 7 天进行考虑，以便你方作出决定。如果你方认为一次短会可以澄清情况，请电话联系我方，以便确定合适的日期与时间来安排会议。

你忠诚的

抄送：工料测量师

Dear Sir

I refer to your application for loss and/or expense [*substitute 'expense' when using GC/Works/1 (1998)*] dated [*insert date*].

Although it appears, at first sight, that you have satisfied all the procedural requirements of the contract and I can begin to consider your application, the presentation is confusing. I am prepared to examine the application as it stands, but I must warn you that it is my duty to reject it, or any part of it, if I am not satisfied of its validity.

Please consider whether you wish to resubmit your application in a more comprehensible form. I will postpone my consideration for 7 days to give you the opportunity to decide. If you think that a short meeting would clarify the situation, I am prepared to arrange it if you will telephone to agree a suitable day and time.

Yours faithfully

Copy: Quantity surveyor

函件 196

致承包商，收到依据补充条款第 S7 条规定的损失和（或）费用估算单

此信函仅适用于 WCD98 合同

Letter 196

To contractor, after receipt of estimate of loss and/or expense under supplementary provision S7

This letter is only suitable for use with WCD 98

尊敬的先生：

谨提及你方提交的合同总额之外的关于直接损失和（或）费用的补充金额估算单，以及你方于［填入日期］发出的中期付款申请书。

［填入：］

我方接受你方的估算单。

［或：］

我方希望就合同总额以外的补充金额进行协商，因双方未同意按第 6A 条/6B 条［视情况取舍］决定此事。

［或：］

与此估算单相关的损失和（或）费用估算按合同第 26 条规定进行。

你忠诚的

Dear Sir

I refer to your estimate of the addition to the contract sum, which you require in respect of direct loss and/or expense, submitted with your application for interim payment dated [*insert date*].

[*Either*:]

I accept your estimate.

[*Or*:]

I wish to negotiate on the amount of the addition to the contract sum and in default of agreement to refer the issue for decision under article 6A/6B [*delete as appropriate*].

[*Or*:]

The provisions of clause 26 shall apply in respect of the loss and/or expense to which the estimate relates.

Yours faithfully

函件 197

致承包商，需要进一步的信息支持以进行费用索赔

此信函不适用于 MW98 合同

Letter 197

To contractor, requesting further information in support of financial claim

This letter is not suitable for use with MW 98

尊敬的先生：

谨提及你方于［填入日期］发出的损失和（或）费用［使用 GC/Works/1 (1998) 合同时，替换为“费用”］申请书，［视情况，加上：］以及我方于［填入日期］收到的进一步的详细说明。

依据第 26.1 条［使用 WCD98 合同时，替换为“第 S7.4 条”；使用 IFC98 合同时，替换为“第 4.11 条”；使用 GC/Works/1（1998）合同时，替换为“第 46(3)(b)条”］，我方在开始考虑你方的申请书之前，需要以下信息：

［清楚地描述所需信息的细节］

你忠诚的

抄送：工料测量师

Dear Sir

I refer to your application for loss and/or expense [*substitute 'expense' when using GC/Works/1 (1998)*] dated [*insert date and add, if appropriate*:] and the further particulars which I received on the [*insert date*].

In accordance with clause 26.1 [*substitute 'S7.4' when using supplementary provisions of WCD 98, '4.11' when using IFC 98 or '46(3)(b)' when using GC/Works/1 (1998)*] I require the following information before I can begin to consider your application:

[*Insert clear details of information required*]

Yours faithfully

Copy: Quantity surveyor

函件 198

致承包商，拒绝损失和（或）费用申请

此信函不适用于 WCD98 合同或 MW98 合同

Letter 198

To contractor, rejecting application for loss and/or expense

This letter is not suitable for use with WCD 98 or MW 98

尊敬的先生：

我方于［填入日期］收到你方按合同第 26 条［使用 IFC98 合同时，替换为“第 4.11 条”；使用 GC/Works/1（1998）合同时，替换为“第 46 条”］发出的损失和（或）费用［使用 GC/Works/1（1998）合同时，替换为“费用”］申请书。

经过仔细考虑，我方必须通知你方，我方认为目前没有任何依据以确定任何损失和（或）费用［使用 GC/Works/1（1998）合同时，替换为“费用”］的发生。

你忠诚的

抄送：工料测量师

Dear Sir

I refer to your application for loss and/or expense [*substitute 'expense' when using GC/Works/1 (1998)*], under clause 26 [*substitute '4. 11' when using IFC 98 or '46' when using GC/Works/1 (1998)*] of the contract, received on the [*insert date*].

After careful consideration of the evidence, I have to inform you that I can find no grounds for ascertaining any loss and/or expense [*substitute 'expense' when using GC/Works/1 (1998)*] at this time.

Yours faithfully

Copy: Quantity surveyor

函件 199

致承包商，接受费用索赔

此信函不适用于 MW98 合同

Letter 199

To contractor, accepting financial claim

This letter is not suitable for use with MW 98

尊敬的先生：

谨提及你方的损失和（或）费用［使用 GC/Works/1（1998）合同时，替换为"费用"］申请书。我方已经于［填入日期］收到此申请书的最后一部分。

我方仔细考虑了这些证据，认为你方的申请书比较合情合理。因此，我方正着手/我方正要求工料测量师［视情况取舍］确定加到合同总额里的补充金额。工料测量师很快会要求你方提供必要的损失和（或）费用的详细情况以确定金额。

你忠诚的

Dear Sir

I refer to your application for loss and/or expense [*substitute* '*expense*' *when using GC/Works/1* (*1998*)], the last part of which I received on [*insert date*].

I have carefully considered the evidence and I am of the opinion that there is some merit in your application. Therefore, I am proceeding/I am asking the quantity surveyor [*delete as appropriate*] to ascertain the amount to be added to the contract sum. He will shortly be requesting you to provide the details of loss and/or expense reasonably necessary for the ascertainment.

Yours faithfully

函件 200

致工料测量师，要求确定损失和（或）费用金额

此信函不适用于 MW98 合同

Letter 200

To quantity surveyor, requesting ascertainment of loss and/or expense

This letter is not suitable for use with MW 98

尊敬的先生：

随函附上承包商寄来的于［填入日期］发出的关于损失和（或）费用［使用 GC/Works/1（1998）合同时，替换为“费用”］的申请书。

我方认为，此申请书完全符合规范，且体现了承包商的有效权利。十分希望你方能尽快着手按实际损失和实际费用来确定应向承包商支付的金额。

你忠诚的

Dear Sir

I enclose an application dated [*insert date*] received from the contractor in respect of loss and/or expense [*substitute 'expense' when using GC/Works/1 (1998)*].

In my opinion, the application is properly made and discloses a valid entitlement. I should be pleased if you would proceed as soon as possible to ascertain the payment due to the contractor, based on actual loss and actual expense.

Yours faithfully

函件 201

致委托人，随函附上关于申请书的报告

此信函不适用于 MW98 合同

Letter 201

To client，enclosing report on application

This letter is not suitable for use with MW 98

尊敬的先生：

我方收到承包商寄来的申请书，其要求支付由于［简要填入事件］引起的损失和（或）费用赔款。

在做关于支付权利的决定时，我必须严格地遵照合同的条款。

随函附上我方的一份简要报告，其中陈述了申请书的主要内容以及我方对每一项内容的决定。此报告供你方参考，如果有内容不清之处，我方将十分乐意做出解释。

你忠诚的

Dear Sir

I have received an application from the contractor in which he asks for payment of loss and/or expense caused by［*insert matters briefly*］.

In making my decision on any entitlement to payment，I must act strictly in accordance with the terms of the contract.

I enclose a copy of my brief report setting out the main points of the application and my decisions in each case. This report is for your information，but I will be happy to explain anything which may seem unclear.

Yours faithfully

函件 202

关于设备的测试和调试，致顾问

Letter 202

To consultants, regarding testing and commissioning of plant

尊敬的先生：

你方计划对设备进行测试和调试的日期为［填入日期］。请回信告知我方此日期是否仍然有效，以便我方邀请委托人到场参加。

你忠诚的

Dear Sir

Your programmed dates for testing and commissioning of [*insert nature of plant*] are [*insert dates*]. Please let me know by return if these dates are still valid so that I can invite the client to be present.

Yours faithfully

函件 203

关于设备的测试和调试，致委托人

Letter 203

To client，regarding testing and commissioning of plant

尊敬的先生：

我方已经安排于［填入日期］［填入时间］开始对［填入设备性质］进行测试和调试。我方建议你方或你方代表到场监督调试程序，以便满意设施的运行情况。

你忠诚的

抄送：顾问
工程监督员

Dear Sir

Arrangements have been made for the［*insert nature of plant*］to be tested and commissioned on the［*insert dates*］beginning at［*insert time*］. It would be advisable for you or your representative to be present to witness the procedure and to satisfy yourself with regard to the operation of the plant.

Yours faithfully

Copies：Consultants
Clerk of works

函件 204

关于收到提前 7 天发出的暂停履行义务的通知，雇主致承包商信函草稿

Letter 204

Draft letter from employer to contractor, on receipt of 7- day notice suspending performance of obligations

尊敬的先生：

谨提及你方注明［填入日期］的作为提前 7 天发出的暂停施工通知的来函。

［如果此通知有效，加上：］

我方确认应付金额将于通知期限到期之前全额支付给你方。

［如果此通知无效，加上：］

此通知包含一个严重错误。

［或：］

经过认真考虑，我方认为你方的通知意思含糊。

［或：］

你方的通知未考虑到我方于［填入日期］发出的扣除通知。

［然后加上：］

因此，你方的通知是无效的并且不产生任何作用。请注意，如果你方继续暂停履行义务，将是不合法的。这样将构成拒绝履行合同，你方需要支付大笔赔偿金。

你忠诚的

Dear Sir

I refer to your letter dated [*insert date*] which purports to be a 7- day suspension notice.

[*If the notice is valid, add*:]

I confirm that payment of the amount due will be made in full before the expiry of the notice period.

[*Otherwise, add either*:]

The purported notice contains a serious error.

[*Or*:]

We are advised that your purported notice is ambiguous.

[*Or*:]

Your purported notice does not take account of my withholding notice dated [*insert date*].

[*Then add*:]

Your notice is, therefore, invalid and of no effect. Take notice that if you proceed to suspend performance of your obligations, it will be unlawful. It will amount to repudiation and you will be liable to me in substantial damages.

Yours faithfully

函件 205

关于收到承包商提前 7 天发出的暂停履行义务的通知，致委托人

通过传真和专递

Letter 205

To client, on receipt of contractor's 7- day notice suspending performance of obligations

By fax and special delivery

尊敬的先生：

我方于今天收到承包商提前 7 天发出的暂停履行义务的通知。如果他方关于你方未支付应付款项之说属实，我方建议你方立即付款，通过快递送去支票并索要收据。

如果你方允许承包商暂停履约，就意味着暂停施工，这样承包商就有权要求工程延期，并申请偿付经济损失。这样也使得承包商有权利暂停对工程的所有投保。

但是，如果你方知道由于某些原因导致他方的通知无效，请立即通知我方，以便我方为你方起草一份回信寄给承包商。

你忠诚的

Dear Sir

I have today received a copy of the contractor's 7-day notice suspending performance of obligations. If he is correct and you have not paid the amount due, may I suggest that you pay without delay, sending the cheque by courier and obtaining a receipt.

If you allow the contractor to suspend, it will not only mean a suspension of building operations for which he will be entitled to an extension of time and to make application for financial losses, he will also be entitled to suspend all insurance of the Works.

If, however, you know of some reason why his notice is invalid, please let me know immediately so that I can draft a reply for you to send.

Yours faithfully

函件 206

关于暂停施工后的复工，致承包商

通过传真和专递

Letter 206

To contractor, on resumption of obligations after suspension

By fax and special delivery

尊敬的先生：

我方注意到你方在暂停施工后已经计划复工，我方对此倍感欣喜。请提供证据证明按合同规定你方有义务投保的所有保险仍然有效。

你忠诚的

Dear Sir

I am glad to see that you propose to resume work on site after your recent suspension. Please provide evidence to demonstrate that all the insurances which you are obliged to maintain under the contract are still in place.

Yours faithfully

函件 207a

致承包商，发出违约通知

此信函仅适用于 JCT98 合同和 WCD98 合同

专递/挂号邮件

Letter 207a

To contractor, giving notice of default

This letter is only suitable for use with JCT 98 and WCD 98

Special/recorded delivery

尊敬的先生：

按照合同条件第 27. 2 条，我方向你方发出通知：你方于以下几方面违约：

[填入违约细节，视情况加上日期]

如果你方在收到此通知后继续违约达 14 天或者你方在收到通知后又重复违约行为，无论你方是否以前重复违约，雇主可以在你方继续违约 10 天内或在你方重复违约行为时再次发出通知，终止此合同对你方的雇佣关系。

你忠诚的

抄送：雇主
工料测量师

Dear Sir

I hereby give you notice under clause 27. 2 of the conditions of contract that you are in default in the following respects:

[*Insert details of the default with dates if appropriate*]

If you continue the default for 14 days after receipt of this notice or if at any time you repeat such default, whether previously repeated or not, the employer may within 10 days of such continuance or repetition by a further notice determine your employment under this contract.

Yours faithfully

Copies: Employer
Quantity surveyor

函件 207b

致承包商，发出违约通知

此信函仅适用于 IFC98 合同

专递/挂号邮件

Letter 207b

To contractor, giving notice of default

This letter is only suitable for use with IFC 98

Special/recorded delivery

尊敬的先生：

按照合同第 7.2.1 条，我方向你方发出通知：你方于以下几方面违约：

[填入违约细节，视情况加上日期]

如果你方在收到此通知后继续违约达 14 天或者你方在收到通知后又重复违约行为，无论你方是否以前重复违约，雇主可以在你方继续违约 10 天内或在你方重复违约行为时再次发出通知，终止此合同对你方的雇佣关系。

你忠诚的

抄送：雇主
工料测量师

Dear Sir

I hereby give notice under clause 7. 2. 1 of the contract that you are in default in the following respects:

[*Insert details of the default with dates if appropriate*]

If you continue the default for 14 days after receipt of this notice or if at any time you repeat such default, whether previously repeated or not, the employer may within 10 days of such continuance or repetition by a further notice determine your employment under this contract.

Yours faithfully

Copies: Employer
Quantity surveyor

函件 207c

致承包商，发出违约通知

此信函仅适用于 MW98 合同

专递/挂号邮件

Letter 207c

To contractor, giving notice of default

This letter is only suitable for use with MW 98

Special/recorded delivery

尊敬的先生：

按照合同第 7.2.1 条，我方向你方发出通知：你方于以下几方面违约：

[填入违约细节，视情况加上日期]

如果你方在收到此通知后继续违约达 7 天，雇主可以再次发出通知，终止此合同对你方的雇佣关系。

你忠诚的

抄送：雇主
工料测量师

Dear Sir

I hereby give notice under clause 7.2.1 of the contract that you are in default in the following respects:

[*Insert details of the default with dates if appropriate*]

If you continue the default for 7 days after receipt of this notice, the employer may thereupon by a further notice determine your employment under this contract.

Yours faithfully

Copies: Employer
Quantity surveyor [*if appointed*]

函件 207d

致承包商，发出违约通知

此信函仅适用于 GC/Works/1（1998）合同

专递/挂号邮件

Letter 207d

To contractor, giving notice of default

This letter is only suitable for use with GC/Works/1 (1998)

Special/recorded delivery

尊敬的先生：

现向你方发出通知，目前存在以下依据可以终止对你方的雇佣关系［按合同第 56(6)(a)，(b)或(e)条以及实际情况，描述终止雇佣的依据］。

如果这些依据至［填入承包商将收到此通知日期 14 天后的日期］仍然存在，雇主将有权再次发出通知，终止此合同对你方的雇佣关系。

你忠诚的

Dear Sir

I hereby give notice that the following grounds for determination exist [*describe grounds for determination under clauses 56(6)(a), (b) or(e) and the facts*].

If these grounds are still in existence on [*insert date 14 days after the date the contractor will receive this notice*], the employer will be entitled to determine the contract forthwith by further notice.

Yours faithfully

函件 208a

如果承包商继续违约，致委托人

此信函不适用于 MW98 合同或 GC/Works/1（1998）合同

Letter 208a

To client, if contractor continues his default

This letter is not suitable for use with MW 98 or GC/Works/1（1998）

尊敬的先生：

谨提及我们于［填入日期］进行的谈话，随后我方向承包商发出了违约通知书。

今晨通知书期限已到，我方在工程监督员陪同下进行了工地检查。我方必须向你方报告［填入违约细节］仍然在继续。按合同条款及特殊条款第 27.2.2 条［使用 IFC98 合同时，替换为“第 7.2.2 条”］，如果你方愿意，你方可以终止与该承包商的雇佣关系，为此我方已随函附上一份可供你方使用的信函草稿。我方预计你方可能会希望在采取进一步行动之前与我方讨论此事，我方将于明日电话联系你方。

请注意，如果你方决定终止与承包商的雇佣关系，你方必须最迟于［填入日期］之前采取行动。

你忠诚的

Dear Sir

I refer to our conversation on the [*insert date*] following which I sent a notice of default to the contractor.

The notice period has expired this morning and I carried out a site inspection in company with the clerk of works. I have to report that [*insert details of the default*] is continuing. Under the terms of the contract and specifically clause 27. 2. 2 [*substitute* '*7. 2. 2*' *when using IFC 98*] you may, if you so wish, determine the employment of the contractor and I enclose a draft letter which you should use for that purpose. I anticipate that you will wish to discuss the matter with me before taking any further action and I will telephone you tomorrow.

Please note that if you decide to determine the contractor's employment, you must do so by [*insert date*] at the latest.

Yours faithfully

函件 208b

如果承包商继续违约，致委托人

此信函仅适用于 MW98 合同

Letter 208b

To client, if contractor continues his default

This letter is only suitable for use with MW 98

尊敬的先生：

谨提及我们于［填入日期］进行的谈话，随后我方向承包商发出了违约通知书。

今晨，我方在工程监督员陪同下进行了工地检查。我方必须向你方报告［填入违约细节］仍然在继续。按合同条款及特殊条款第 7.2.1 条，如果你方愿意，你方可以终止与该承包商的雇佣关系，为此我方已随函附上一份可供你方使用的信函草稿。我方预计你方可能会希望在采取进一步行动之前与我方讨论此事，我方将于明日电话联系你方。

请注意，如果你方决定终止与承包商的雇佣关系，你方必须立即采取行动。

你忠诚的

Dear Sir

I refer to our conversation on the [*insert date*] following which I sent a notice of default to the contractor.

I carried out a site inspection this morning in company with the clerk of works and I have to report that [*insert details of the default*] is continuing. Under the terms of the contract and specifically clause 7.2.1 you may, if you so wish, determine the employment of the contractor and I enclose a draft letter which you should use for that purpose. I anticipate that you will wish to discuss the matter with me before taking any further action and I will telephone you tomorrow.

Please note that if you decide to determine the contractor's employment, you must not delay in taking action.

Yours faithfully

函件 208c

如果承包商违约，致委托人

此信函仅适用于 GC/Works/1（1998）合同

Letter 208c

To client, if contractor is in default

This letter is only suitable for use with GC/Works/1（1998）

尊敬的先生：

谨提及我们于［填入日期］讨论了［简要填入承包商的违约细节］。随后，按我们的一致意见，我方向承包商发出了违约通知书[1]。

今晨，我方在工程监督员陪同下进行了工地检查。我方必须向你方报告［填入违约细节］在承包商收到违约信函后，已经继续了 14 天。很明显，情况非常严重，我方相信承包商并未就违约之处进行改正。按合同条款及特殊条款第 56 条，如果你方愿意，你方可以终止与该承包商的雇佣关系，为此我方已随函附上一份可供你方使用的信函草稿。我方预计你方可能会希望在采取进一步行动之前与我方讨论此事，我方将于明日电话联系你方。

你忠诚的

［1 此信函应该在承包商收到违约通知 14 天后发出。］

Dear Sir

I refer to our conversation on the [*insert date*] when we discussed [*insert brief details of the contractor's default*]. Following that conversation, as agreed, I issued a default notice to the contractor[1].

I carried out a site inspection this morning in company with the clerk of works and I have to report that [*insert details of the default*] is continuing 14 days after the default letter. The situation is clearly very serious and I believe that the contractor is making little effort to rectify matters. Under the terms of the contract and specifically clause 56 you may, if you so wish, determine the contract forthwith and I enclose a draft letter which you could use for that purpose. I anticipate that you will wish to discuss the matter with me before taking any further action and I will telephone you tomorrow.

Yours faithfully

[1 *This letter should not be sent until 14 days after the date the contractor has received the default letter.*]

函件 208d

如果按合同第 28A 条或 7. 13. 1 条有可能终止合同，致委托人

此信函不适用于 MW98 合同或 GC/Works/1（1998）合同

专递/挂号邮件

Letter 208d

To client, if determination under 28A or 7. 13. 1 possible

This letter is not suitable for use with MW 98 or GC/Works/1 (1998)

Special/recorded delivery

尊敬的先生：

全部工程或大部分工程将于［填入日期］暂停施工［填入附件中规定的期限］。之后，你方可以立即发出一份通知：除非承包商在收到通知后 7 天内结束停工，否则其雇佣关系将被终止。这是后果很严重的一步行动，如果你方愿意，我方可以为你方起草一份适当的信函。不过，我方建议与你方面议，详细讨论其可能的后果。我方会在几天内与你方电话联系。

你忠诚的

Dear Sir

On the [*insert date*] the whole or substantially the whole of the Works will have been suspended for [*insert period stated in the appendix*]. Immediately thereafter, you may issue a notice that unless the suspension is terminated within 7 days of receipt of the notice, the contractor's employment under the contract will determine. This is an extremely serious step and if you wish I can draft you an appropriate notice. However, I suggest that we should meet to discuss the possible consequences in detail. I will telephone you in the next few days.

Yours faithfully

函件 209a

雇主致承包商的函件草稿，终止雇佣关系

此信函不适用于 MW98 合同或 GC/Works/1（1998）合同

专递/挂号邮件

Letter 209a

Draft letter from employer to contractor, determining employment

This letter is not suitable for use with MW 98 or GC/Works/1（*1998*）

Special/recorded delivery

尊敬的先生：

谨提及建筑师［使用 WCD98 合同时，省略］寄给你方的于［填入日期］发出的通知。

依据合同第 27.2.2 条［使用 IFC98 合同时，替换为“第 7.2.2 条”］，请把此函作为正式通知，我方终止此合同对你方的雇佣关系，但不影响我方拥有的任何其他权利或索赔要求。

双方的权利和义务须遵循合同第 27.6 条和第 27.7 条［使用 IFC98 合同时，替换为“第 7.6 条和第 7.7 条”］的规定。在建筑师发出指示之前，不得拆除任何临时建筑，或运走任何施工设施、工具、设备、货物及材料。

建筑师会在 14 天内就所有分包商或供应商的事项致函你方［使用 JCT 98 合同时，加上：］指定的或非指定的。

你忠诚的

抄送：建筑师

工料测量师

Dear Sir

I refer to the notice dated [*insert date of notice*] sent to you by the architect [*omit if using WCD 98*].

In accordance with clause 27.2.2 [*substitute '7.2.2' when using IFC 98*] of the contract take this as notice that I hereby determine your employment under this contract without prejudice to any other rights or remedies which I may possess.

The rights and duties of the parties are governed by clauses 27.6 and 27.7 [*substitute '7.6 and 7.7' when using IFC 98*]. No temporary buildings, plant, tools, equipment, goods or materials shall be removed from site until and if the architect shall so instruct.

The architect will write to you within 14 days regarding all sub-contractors and suppliers [*when using JCT 98, add:*] whether nominated or otherwise.

Yours faithfully

Copies: Architect
Quantity surveyor

函件 209b

雇主致承包商的函件草稿，发生损失或损害后，终止雇佣关系

此信函不适用于 MW98 合同或 GC/Works/1（1998）合同

专递/挂号邮件

Letter 209b

Draft letter from employer to contractor, determining employment after loss or damage

This letter is not suitable for use with MW 98 or GC/Works/1 (1998)

Special/recorded delivery

尊敬的先生：

谨提及你方于［填入日期］寄来的关于［填入由于已经投保的某种或几种危险导致的损失或损害细节］损失或损害的通知。

依据合同第 22C. 4. 3. 1 条［使用 IFC98 合同时，替换为“第 6. 3C. 4. 3 条”］，请把此函作为正式通知，我方终止就此合同对你方的雇佣关系，因为我方认为这样做是公正而且公平的。

双方的权利和义务须遵循合同第 22C. 4. 3. 2 条［使用 IFC98 合同时，替换为“第 6. 3C. 4. 4 条”］的规定。

你忠诚的

抄送：建筑师

工料测量师

Dear Sir

I refer to your notice of the [*insert date*] giving notice of loss or damage occasioned by [*insert particulars of loss or damage which must have been occasioned by one or more of the insured risks*].

In accordance with clause 22C. 4. 3. 1 [*substitute '6. 3C. 4. 3' when using IFC 98*] take this as notice that I hereby determine your employment under the contract, because I consider that it is just and equitable so to do.

The rights and duties of the parties are governed by clause 22C. 4. 3. 2 [*substitute '6. 3C. 4. 4' when using IFC 98*].

Yours faithfully

Copies: Architect
Quantity surveyor

函件 209c

雇主致承包商的函件草稿，按合同第 28A 条或 7. 13. 1 条终止雇佣关系

此信函不适用于 MW98 合同或 GC/Works/1 （1998） 合同

专递/挂号邮件

Letter 209c

Draft letter from employer to contractor, determining employment under clause 28A or 7. 13. 1

This letter is not suitable for use with MW 98 or GC/Works/1 (1998)

Special/recorded delivery

尊敬的先生：

由于［填入暂时停工的原因］，全部工程或大部分工程已经于［填入日期］起，暂停施工达［填入附件中规定的期限］。

依据合同第 28A. 1. 1 条［使用 IFC98 合同时，替换为“第 7. 13. 1 条”］，请把此函作为通知：如果你方在收到此信函后 7 天内没有终止停工，对你方就此合同的雇佣关系将终止。

双方的权利和义务须遵循合同第 28A. 2 条至第 28A. 7 条［使用 IFC98 合同时，替换为“第 7. 14 条至第 7. 19 条”］。一旦有可能，我方将立刻作出账目报表。

你忠诚的

抄送：建筑师

工料测量师

Dear Sir

The whole or substantially the whole of the Works has been suspended since [*insert date*], a period of [*insert length stated in the appendix*], by reason of [*insert reason for suspension*].

In accordance with clause 28A. 1. 1 [*substitute '7. 13. 1' when using IFC 98*] of the contract, take this as notice that unless the suspension is terminated within 7 days of the date of receipt of this notice, your employment under this contract will determine 7 days after receipt of this notice.

The rights and duties of the parties are governed by clauses 28A. 2 to 28A. 7 [*substitute '7. 14 to 7. 19' when using IFC 98*]. I will draw up a statement of account as soon as reasonably practicable.

Yours faithfully

Copies: Architect
Quantity surveyor

函件 209d

雇主致承包商的函件草稿，终止雇佣关系

此信函仅适用于 MW98 合同

专递/挂号邮件

Letter 209d

Draft letter from employer to contractor, determining employment

This letter is only suitable for use with MW 98

Special/recorded delivery

尊敬的先生：

谨提及建筑师寄给你方的于［填入通知日期］发出的通知。

依据合同第 7.2.1 条，请把此函作为正式通知：我方终止就此合同对你方的雇佣关系，但不影响我方拥有的任何其他权利或索赔要求。

你方必须立即撤出工地。

双方的权利和义务须遵循合同第 7.2.3 条的规定。请注意，在工程竣工及缺陷整改完成之前，我方无须继续向你方付款。我方将保留此权利到竣工时。

你忠诚的

抄送：建筑师
工料测量师

Dear Sir

I refer to the notice dated [*insert date of notice*] sent to you by the architect.

In accordance with clause 7. 2. 1 of the contract take this as notice that I hereby forthwith determine your employment under this contract without prejudice to any other rights or remedies which I may possess.

You must immediately give up possession of the site.

The rights and duties of the parties are governed by clause 7. 2. 3. Take note that I am not bound to make any further payments to you until after completion of the Works and the making good of any defects therein. I reserve any rights to that time.

Yours faithfully

Copies: Architect
Quantity surveyor [*if appointed*]

函件 209e

雇主致承包商的函件草稿，终止合同

此信函仅适用于 GC/Works/1（1998）合同

专递/挂号邮件

Letter 209e

Draft letter from employer to contractor, determining contract

This letter is only suitable for use with GC/Works/1 (1998)

Special/recorded delivery

尊敬的先生：

依据合同第 56(1)(a)条，我方就此发出正式通知终止合同。

[视情况，加上：]

以下为终止合同的依据：[描述属于合同第 56(6)(a)、(b)或(e)条的任何依据]。

[然后：]

双方的权利和义务须遵循合同第 57 条的规定。一旦有可能，我方将立即致函你方，无论如何会于 3 个月之内/竣工之前 [视情况取舍]，按第 56(3)条发出指示。在我方发出指示之前，你方不能拆除或搬走任何东西。

你忠诚的

抄送：建筑师

工料测量师

Dear Sir

Take this as notice under clause 56(1)(a) of the contract that I hereby determine this contract.

[*If appropriate add*:]

The following grounds apply: [*describe any grounds which fall within clause 56(6)(a), (b) or(e)*]

[*Then*:]

The rights of the parties are governed by clause 57. I will write to you as soon as practicable, but in any event not later than 3 months hence/the date of completion [*delete whichever is later*] to give directions under clause 56(3). You must not remove any Things of whatsoever kind before you receive my directions.

Yours faithfully

Copies: Architect
Quantity surveyor

函件 210

关于终止合同后的保险，致委托人

Letter 210

To client，regarding insurance after determination

尊敬的先生：

你方的终止合同的通知已经于今天寄给承包商。承包商不再承担对工程投保的责任。你方应该立即与保险经纪人商讨，以获得与前承包商按合同条款所投保的类似的保险。现随函附上合同第20条到第22A条［使用IFC98合同或MW98合同时，替换为“第6.1条到第6.3A条”；或使用GC/Works/1（1998）合同时，替换为“第8条”］的副本数份，请转交你方经纪人。

你方应持续投保，直到安排好雇佣其他承包商来完成工程施工。

你忠诚的

抄送：工料测量师

Dear Sir

Your notice of determination is being sent to the contractor today. The contractor no longer has any liability to insure the works. You should consult your own broker without delay to obtain cover similar to that which the contractor was required to have under the provisions of this contract. I enclose copies of clauses 20 to 22A [*substitute* '*6.1 to 6.3A*' *when using IFC 98 or MW98 or* '*8*' *when using GC/Works/1* (*1998*)] which you should give to your broker.

Your insurance cover should be maintained at least until suitable arrangements have been made to complete the Works using another contractor.

Yours faithfully

Copy：Quantity surveyor

函件 211

若承包商可能会终止合同雇佣关系，致委托人

此信函不适用于 GC/Works/1（1998）合同

Letter 211

To client, if contractor likely to determine his employment under the contract

This letter is not suitable for use with GC/Works/1 (1998)

尊敬的先生：

我方有理由相信，承包商正在严肃考虑采取行动终止合同的雇佣关系。

这样做在时间和资金方面将对此项目造成破坏性的后果。我方已经在我处安排与承包商召开专门会议，听取承包商的苦处。会议定于［填入日期］［填入时间］进行。如果你方能随时准备批准任何可能需要采取的行动，我方将不胜感激。无论如何，我方将于会后立即电话通知你方结果。

你忠诚的

Dear Sir

I have reason to believe that the contractor is seriously considering taking action to determine his employment under the contract.

The results of such action would be disastrous for the project in terms of time and money and I have arranged a special meeting with the contractor at this office to hear his grievances. The meeting will be at [*insert time*] on [*insert date*] and I should be grateful if you would keep yourself available to approve any action which may need to be taken. In any event, I will telephone immediately after the meeting to inform you of the result.

Yours faithfully

函件 212

如果承包商终止雇佣关系，致委托人

此信函不适用于 GC/Works/1（1998）合同

Letter 212

To client, if contractor determines his employment

This letter is not suitable for use with GC/Works/1（1998）

尊敬的先生：

我方得知承包商于今日寄来通知书，终止了此合同的雇佣关系。

这样做会导致很严重的后果。你方应该紧急召集会议征询专家意见。我方将和工料测量师一起陪同你方到会，以便提供必需的信息。我们将随时待命。

你忠诚的

抄送：工料测量师

Dear Sir

I understand that the contractor has today sent notice determining his employment under the contract.

The result of this action could be very serious and you should take specialist advice as a matter of urgency. I will accompany you at the meeting together with the quantity surveyor in order to provide any information that may be required. We will make ourselves available at any time.

Yours faithfully

Copy: Quantity surveyor

函件 213

雇主致承包商的信函草稿，发出拟将争端事项诉诸裁决的通知

专递邮件

Letter 213

Draft letter from employer to contractor, giving notice of intention to refer a dispute to adjudication

Special delivery

尊敬的先生：

依据合同第 41A. 4. 1 条［使用 WCD98 合同时，替换为“第 39A. 4. 1 条”；使用 IFC98 合同时，替换为“第 9A. 4. 1 条”；使用 MW98 合同时，替换为“第 D4. 1 条”；使用 GC/Works/1（1998）合同时，替换为“第 59（1）条”］，我方打算将以下争端或分歧诉诸裁决：[1]［填入争端内容］。我方将邀请裁决人［填入寻求补偿的内容，如宣布全部粉刷工作或裁决人决定的任何部分不符合合同要求。］

［使用 GC/Works/1（1998）合同时，可添加以下内容：］

裁决人［填入姓名］将为专用条款中规定的人员。

你忠诚的

抄送：指定机构［使用 GC/Works/1（1998）合同除外，因为此合同中裁决人将会获得一份副本］

［[1] 请注意由雇主亲自处理裁决的情况很少，通常他会指示精通起诉判决事项的专人代为处理。］

Dear Sir

Under the provisions of clause 41A. 4. 1 [*substitute '39A. 4. 1' when using WCD 98, '9A. 4. 1' when using IFC 98, 'D4. 1' when using MW 98 or '59 (1)' when using GC/Works/1 (1998)*] we intend to refer the following dispute or difference to adjudication[1]: [*insert a description of the dispute*]. We will be requesting the adjudicator to [*insert the nature of the redress sought, e. g. 'give a declaration that the whole of the paintwork, or such parts as the adjudicator shall decide, is not in accordance with the contract'*].

[*When using GC/Works/1 (1998), add:*]

The adjudicator will be [*insert name*] as specified in the abstract of particulars.

Yours faithfully

Copy: Nominating Body [*except in the case of GC/Works/1 (1998) where the adjudicator should be sent a copy*]

[1 *It should be noted that it is comparatively rare for the employer to deal with an adjudication himself and he will usually instruct someone experienced in adjudication to deal with it on his behalf.*]

函件 214

雇主致指定机构的信函草稿，要求指定裁决人

此信函不适用于 GC/Works/1（1998）合同

专递邮件

Letter 214

Draft letter from employer to Nominating Body, requesting nomination of an adjudicator

This letter is not suitable for use with GC/Works/1 (1998)

Special delivery

尊敬的先生：

现随函附上一份拟将本合同内的争端或分歧诉诸裁决[1]的通知书。我方今天已将本通知书及随函送达合同另一方：[填入承包商姓名]。本合同是以 1998 年 JCT 建筑合同标准格式为依据［使用 WCD98 合同时，为“JCT 承包商设计建筑合同标准格式”；［使用 IFC98 合同时，为“JCT 中型建筑合同标准格式”或使用 MW98 合同时，为“1998 年 JCT 小型建筑工程协议书”］。你方为选定裁决人指定机构。因此，根据合同条件第 41A. 2 条［使用 WCD98 合同时，替换为“第 39A. 2 条”；使用 IFC98 合同时，为“第 9A. 2 条”或使用 MW98 合同时，为“第 D2 条”］规定，我方向贵机构提出指定裁决人的申请。随函还附上了已经填好的申请表副本一份以及索赔人提交的金额为［填入金额］的支票一张。

你忠诚的

抄送：承包商

[1 请注意由雇主亲自处理裁决的情况很少，通常他会指示精通起诉判决事项的专人代为处理。]

Dear Sir

We enclose a notice of intention to refer disputes and/or differences under the contract to adjudication[1]. We have today served this notice and covering letter on the other party to the contract: [*insert the name of the contractor*]. The contract was executed on JCT Standard Form of Building Contract 1998 [*substitute* '*JCT Standard Form of Building Contract With Contractor's Design*' *when using WCD 98*, '*JCT Intermediate Form of Building Contract*' *when using IFC 98 or* '*JCT Agreement for Minor Building Works 1998*' *when using MW 98*]. You are the selected nominator. Therefore, in accordance with clause 41A.2 [*substitute* '*39A.2*' *when using WCD 98*, '*9A.2*' *when using IFC 98 or* '*D2*' *when using MW 98*], we hereby make application to you to appoint an adjudicator. A copy of the completed application form and the claimant's cheque in the sum of [*insert the amount*] is enclosed.

Yours faithfully

Copy: Contractor

[[1] *It should be noted that it is comparatively rare for the employer to deal with an adjudication himself and he will usually instruct someone experienced in adjudication to deal with it on his behalf.*]

函件 215

雇主致裁决人信函草稿，附上提交裁决的争端事项

专递邮件

Letter 215

Draft letter from employer to adjudicator, enclosing the Referral

Special delivery

尊敬的先生：

按照你方的指示，［如果裁决人到雇主希望提交争端事项之日仍未发出任何指示，则省略］我方随函附上此争端事项的文件，以及拟将之诉诸裁决的通知书副本[1]。

此信函副本已寄给被起诉人。

你忠诚的

抄送：承包商

［1 请注意由雇主亲自处理裁决的情况很少，通常他会指示精通起诉判决事项的专人代为处理。］

Dear Sir

In accordance with your directions, [*delete if the adjudicator has not issued any directions at the date when the employer wishes to submit the Referral*] I enclose the Referral documentation in this matter together with a further copy of the notice of intention to refer to adjudication[1].

A copy of this letter has been sent to the Respondent.

Yours faithfully

Copy: Contractor

[1 *It should be noted that it is comparatively rare for the employer to deal with an adjudication himself and he will usually instruct someone experienced in adjudication to deal with it on his behalf.*]

函件 216

如果裁决人的决议有利于雇主，雇主致承包商的信函草稿

专递邮件

Letter 216

Draft letter from employer to contractor, if adjudicator's decision is in employer's favour

Special delivery

尊敬的先生：

毫无疑问，你方已经收到裁决人的决议。

裁决人决议你方必须在［填入日期］之前［填入决议内容，如支付一定金额或进行某种任务］。请确认你方将服从此决议。

如果你方在［填入日期］之前没有［填入决议内容，如支付一定金额或进行某种任务］，我方将立即指示律师采取强制性措施。

你忠诚的

Dear Sir

No doubt you have received the adjudicator's decision.

He has decided that you must [*insert the element of the decision, e. g. pay £ ... or perform some task*] by [*insert date*]. Please confirm that you intend to comply with the decision.

If you have not [*insert the element of the decision, e. g. paid £ ... or performed the task*] by [*insert date*], I will immediately instruct solicitors to commence enforcement proceedings.

Yours faithfully

函件 217

如果被要求对承包商的申请裁决的通知做出回应，致雇主

通过传真和邮寄

Letter 217

To employer, if asked to respond to contractor's adjudication notice.

By fax and post

尊敬的先生:

你方于［填入日期］的来函收悉，深表谢意。信中要求我方代表你方对承包商的申请裁决通知做出回应。

很遗憾，你方所要求的事项已超出我方的能力，因而不在我方的服务范围之内。

你方应该立即指示对此类事件经验丰富的专人或公司来代理此事。由于裁决程序的安排非常之快，请抓紧时间。

你忠诚的

Dear Sir

Thank you for your letter dated [*insert date*] asking me to act for you in responding to the contractor's adjudication notice.

Unfortunately, what you ask is beyond the scope of my expertise and, for that reason, it is not included in my services.

You should immediately instruct a person or firm, experienced in these matters, to represent you. There is no time to lose because the adjudication procedure is designed to be very fast.

Yours faithfully

函件 218a

雇主致承包商的信函草稿，要求共同委任仲裁人

此信函不适用于 GC/Works/1（1998）合同

专递/挂号邮件

Letter 218a

Draft letter from employer to contractor, requesting concurrence in the appointment of an arbitrator

This letter is not suitable for use with GC/Works/1（1998）

Special/recorded delivery

尊敬的先生：

现告知你方，依据我们双方于［填入日期］签订的合同条件第7A 条和第41B 条［使用 WCD98 合同时，替换为“第 6A 条和第 39B 条”；使用 IFC98 合同时，替换为“第 9A 条和第 9B 条”；使用 MW98 合同时，替换为“第 7A 条和第 8.2 条”］的规定，要求以下提到的双方之间的争端或分歧诉诸仲裁。请将本函视为需要共同任命仲裁人的要求。

提交仲裁处理的争端或分歧为：［简要陈述］。

我方建议的下列三位人选供你方考虑，并且要求你方在本函送达之日起 14 天内共同完成任命工作。如若上述三人双方不能共同任命，我方将向英国皇家建筑师协会/皇家特许测量师学会/特许仲裁人学会/法律学会/苏格兰法律学会/土木工程师学会［视情况取舍］的主席或副主席提请任命。

［列出三位人选的姓名和地址］。

你忠诚的

Dear Sir

I hereby give you notice that I require the undermentioned dispute or difference between us to be referred to arbitration in accordance with article 7A and clause 41B [*substitute* '*article 6A and clause 39B*' *when using WCD 98*, '*article 9A and clause 9B*' *when using IFC 98*, '*article 7A and clause 8.2*' *when using MW 98*] of the contract between us dated [*insert date*]. Please treat this as a request to concur in the appointment of an arbitrator.

The dispute or difference is [*insert brief description*].

I propose the following three persons for your consideration and require your concurrence in the appointment within 14 days of the date of service of this letter, failing which I shall apply to the President or Vice-President of the Royal Institute of British Architects/Royal Institution of Chartered Surveyors/Chartered Institute of Arbitrators/Law Society/Law Society of Scotland/Institution of Civil Engineers [*delete as appropriate*].

[*List names and addresses of the three persons*]

Yours faithfully

函件 218b

雇主致承包商的信函草稿，要求共同委任仲裁人

此信函仅适用于GC/Works/1（1998）合同

专递/挂号邮件

Letter 218b

Draft letter from employer to contractor, requesting concurrence in the appointment of an arbitrator

This letter is only suitable for use with GC/Works/1 (1998)

Special/recorded delivery

尊敬的先生：

现告知你方，依据我们双方于［填入日期］签订的合同条件第60条的规定，要求下列提到的双方之间的争端或分歧诉诸仲裁。

提交仲裁处理的争端或分歧为：［简要陈述］。

你忠诚的

Dear Sir

I hereby give you notice that I require the undermentioned dispute or difference between us to be referred to arbitration in accordance with clause 60 of the contract between us dated [*insert date*].

The dispute or difference is [*insert brief description*].

Yours faithfully

函件 219

如果不能达成共同委任仲裁人或 GC/Works/1（1998）合同中的指定仲裁人不能实施仲裁，雇主致专业机构的信函草稿

专递/挂号邮件

Letter 219

Draft letter from employer to Professional Body, if there is no concurrence in the appointment of an arbitrator or, in the case of GC/Works/1 (1998), the named arbitrator is unable to act

Special/recorded delivery

尊敬的先生：

我方是采用 JCT98 合同格式、合同第 41B. 1. 1 条［视情况分别替换为“WCD98 合同格式、第 39B. 1. 1 条”；“IFC98 合同格式、第 9B. 1. 1 条”；“MW98 合同格式、补充条件第 E2. 1 条”，或“GC/Works/1（1998）合同格式、第 60（1）条”的雇主，该合同规定由你方的主席或副主席指定仲裁人。

请你方寄发申请表和有关文件，同时告知申请所需的费用，我方将对此不胜感激。

你忠诚的

Dear Sir

I am an employer who has entered into a building contract in JCT 98 form, clause 41B. 1. 1 [*substitute 'WCD 98 form, clause 39B. 1. 1' 'IFC 98 form, clause 9B. 1. 1', 'MW 98 form, Supplemental Condition E2. 1' or 'GC/Works/1 (1998) form, clause 60 (1)' as appropriate*] which makes provision for your President or Vice-President to appoint an arbitrator.

I should be pleased to receive an appropriate form of application and supporting documentation, together with a note of the current fee payable on application.

Yours faithfully

函件 220

如果某项服务无法执行，致委托人

此信函不适用于 SW/99 合同

专递/挂号邮件

Letter 220

To client, if impractical to carry out certain services

This letter is not suitable for use with SW/99

Special/recorded delivery

尊敬的先生：

按照雇佣条款第 1.6 条，我方必须通知你方，目前出现了以下情况，即［填入简要细节］，因而无法执行［填入服务细节］。

如果你方能紧急电话联系我方，以商定针对这些状况应采取的合适的行动，我方将不胜感激。

你忠诚的

Dear Sir

In accordance with clause 1.6 of the terms of engagement, I am required to give you this notice that a circumstance has arisen, namely [*insert concise details*], which make it impracticable to carry out [*insert details of services*].

I should be grateful, therefore, if you would telephone me as a matter of urgency in order to agree a suitable course of action having regard to all the circumstances.

Yours faithfully

函件 221a

致委托人，提前 14 天发出通知终止履行所有服务和义务

此信函不适用于 SW/99 合同

专递/挂号邮件

Letter 221a

To client, determining all performance and obligations by 14-day notice

This letter is not suitable for use with SW/99

Special/recorded delivery

尊敬的先生：

按雇佣条款第 8.5 条，请将此函作为正式通知，我方打算于［填入终止雇佣关系日期，此日期必须于 14 天后］终止履行所有服务和合同第二部分规定的我方义务。

我方终止履约的依据是：［简要描述］。

我方很乐意与你方面谈因终止履约而引起的各种事项，包括已经准备好的工程图纸及文件的使用、我方的收费，以及委任其他建筑师等。请你方告知方便来我方办公室的日期及时间，或如果你方愿意，请告知我方去你处的合适时间。

你忠诚的

Dear Sir

In accordance with clause 8.5 of the terms of engagement, please take this as notice that I intend to determine performance of all of the Services and my obligations under part 2 of the conditions on the [*insert date of termination*, *which must be after the expiry of 14 days*].

My grounds for determination are [*insert a brief description*].

I shall be happy to meet you to discuss matters arising from this termination, including the use of drawings and documents already prepared, my fees and the appointment of another architect. Please inform me of a date and time which will be convenient for you to attend this office or alternatively, if you prefer, for me to visit you.

Yours faithfully

函件 221b

致委托人，发出合理通知终止履约

此信函仅适用于 SW/99 合同

专递/挂号邮件

Letter 221b

To client, terminating performance by reasonable notice

This letter is only suitable for use with SW/99

Special/recorded delivery

尊敬的先生：

按雇佣条款第 34 条，请将此函作为正式通知，我方打算于［填入终止雇佣关系日期，此日期必须合理考虑到工程的规模、复杂性及目前的阶段］终止履行所有的服务。

我方很乐意与你方面谈因终止履约而引起的各种事项，包括已经准备好的工程图纸及文件的使用、我方的收费，以及委任其他建筑师等。请你方告知方便来我方办公室的日期及时间，或如果你方愿意，请告知我方去你处的合适时间。

你忠诚的

Dear Sir

In accordance with clause 34 of the terms of engagement, please take this as notice that I intend to terminate performance of all of the Services on the [*insert date of termination, which must be reasonable having regard to the size and complexity of the project and the stage reached*].

I shall be happy to meet you to discuss matters arising from this termination, including the use of drawings and documents already prepared, my fees and the appointment of another architect. Please inform me of a date and time which will be convenient for you to attend this office or alternatively, if you prefer, for me to visit you.

Yours faithfully

函件 222

如果委托人发出合理通知终止委任关系，致委托人

专递/挂号邮件

Letter 222

To client, if client terminates appointment by reasonable notice

Special/recorded delivery

尊敬的先生：

你方于［填入日期］的来函收悉，深表谢意。从信中，我方得知你方打算于［填入日期］终止就以上项目对我方作为建筑师的委任关系。

我方正安排于你方说明的日期停止我方的工作，随后我方会立即提交一份包括所有未付费用的账单。你方只有在付清所有款项后，才能有权取得所有本项目涉及的工程图纸和文件资料。尽管这样会造成信息不完整，但我方对于工程错误或疏漏不承担任何责任。

［如果已经完成了 D 阶段，或你方正在索取全额费用，可以加上：］

你方可以复制与此项目相关的设计图纸，但所有图纸和文件的著作权属我方所有。

［如果尚未完成阶段 D，或你方正在索取少量费用，则替换为：］

在不经我方同意的情况下，你方无权复制我方的设计图纸。如果你方额外支付［填入金额］，我方可以允许你方复制与此项目相关的设计图纸。但工程文件的著作权属我方所有。

你忠诚的

Dear Sir

Thank you for your letter of the [*insert date*], by which I understand that you intend to terminate my appointment as architect for the above project on the [*insert date*].

I am arranging to cease my work on the date stated and immediately thereafter I will submit an account to cover all outstanding fees. When I receive your payment, you will be entitled to all the drawings and documents prepared for the work although, in view of the circumstances, the information will be incomplete and I cannot accept responsibility for errors or omissions.

[*If stage D has been completed or if you are charging full fees, add*:]

You are entitled to reproduce the design on the site to which it relates. The copyright in all drawings and documents remains my property.

[*If stage D has not been completed or if you are charging a nominal fee, add instead*:]

You are not entitled to reproduce my designs by executing the project without my permission. I am prepared to grant you permission to reproduce my design on the site to which it relates on payment of an additional fee of [*insert amount*]. The copyright in all the documents remains my property.

Yours faithfully

函件 223a

如果委托人没有在6个月内发出指示，要求恢复暂停的服务，致委托人

此信函不适用于SW/99合同

专递/挂号邮件

Letter 223a

To client, if client has not given instructions to resume suspended service within 6 months

This letter is not suitable for use with SW/99

Special/recorded delivery

尊敬的先生：

你方于［填入日期］发出通知，暂停我方就以上项目的服务，暂停服务的有效日期为［填入日期］。

目前距暂停服务之日已经超过6个月，依据雇佣条款第8.3条，我方特此书面要求你方发出指示，恢复我方的服务。此指示应被视为书面指示。

如果我方在此信函发出后30天内没有收到回复服务的指示，我方有权认为委任关系终止。

你忠诚的

Dear Sir

You suspended my services in connection with the above project by your notice of the [*insert date*]. The effective date of suspension of services was [*insert date*].

It is now more than 6 months from the date of suspension and, in accordance with clause 8.3 of the terms of engagement, I hereby make written request for instructions to resume my services, such instructions to be in writing.

If such instructions have not been received by me within 30 days of the date of this letter, I have the right to treat the appointment as terminated.

Yours faithfully

函件 223b

如果委托人没有在6个月内发出指示，要求恢复暂停的服务，致委托人

此信函仅适用于SW/99合同

专递/挂号邮件

Letter 223b

To client, if client has not given instructions to resume suspended service within 6 months

This letter is only suitable for use with SW/99

Special/recorded delivery

尊敬的先生：

你方于［填入日期］发出通知，暂停我方就以上项目的服务，暂停服务的有效日期为［填入日期］。

目前已经超过暂停服务之日6个月，依据雇佣条款第34条，我方就此终止委任关系。

你忠诚的

Dear Sir

You suspended my services in connection with the above project by your notice of the [*insert date*]. The effective date of suspension of services was [*insert date*].

It is now more than 6 months from the date of suspension and, in accordance with clause 34 of the terms of engagement, I hereby terminate the Appointment.

Yours faithfully

函件 224

如果委任其他建筑师，致委托人

专递/挂号邮件

Letter 224

To client, if another architect appointed

Special/recorded delivery

尊敬的先生：

听闻你方已委任了另外一位建筑师进行以上项目，我方颇为吃惊。请你方确认此消息是否属实。

我方的委任关系将会持续到你方按雇佣条款第 8.5 条［使用 SW/99 合同时，替换为“第 34 条”］正式发出合理的终止委任关系的通知为止。欣然希望你方能对此事给予说明，以便我方采取适当的行动，包括停止工作、封存文档，以及做出费用账单。

你忠诚的

Dear Sir

I was surprised to hear that you have appointed another architect to carry out the above project. Perhaps you would be good enough to confirm that my information is correct?

My own appointment, of course, continues until you formally give me reasonable notice of termination in accordance with clause 8.5 [*substitute '34' when using SW/99*] of the terms of engagement. I should be pleased to hear from you on this matter so I can take appropriate action, including stopping work, closing my files and preparing my fee account.

Yours faithfully

函件 225

致委托人委任的其他建筑师

专递/挂号邮件

Letter 225

To other architect, appointed by client

Special/recorded delivery

尊敬的先生：

我方得知我方的委托人［填入姓名］指示你方来进行以上项目。由于我方没有收到任何你方按照 RIBA 的《职业操守规范》第 3 则第 3.8 条规定发出的信函，我方欣然希望你方能告知此消息是否正确。

很明显，你方并不知道我方之前在负责此项目，但如果你方接受了我方委托人的指示，我方建议你方应该向其说明，在办完所有包括我方的费用支付在内的相应的终止手续之前，你方不应该继续该项目。

我方已经致函给委托人，告知其要严格按照雇佣条款的规定对我方进行必要的终止委任手续。

你忠诚的

Dear Sir

I have been informed that my client [*insert name*] has instructed you to carry out work on the above project. Since I have not had any communication from you in accordance with principle 3, rule 3.8 of the RIBA Code of Professional Conduct, I should be pleased if you would let me know if my information is correct.

Clearly, you were not aware of my prior involvement, but I suggest that if you have accepted my client's instructions, you should inform him that it would not be proper for you to proceed until the appropriate termination formalities, including payment of my fees, have been completed.

I have written to my client informing him of the necessity of terminating my appointment in strict accordance with the applicable conditions of my terms of engagement.

Yours faithfully

第十一章　工程竣工

这些信函内容涉及为工程竣工之前的检查所作的准备，缺陷责任期以及发放最终证明书之前的相关事宜。

本书没有提供证书样本，因为为特定目的印刷的规范表格可以很方便地找到。

函件 226

如果委托人希望在证明竣工之前使用新建建筑物，致委托人

Letter 226

To client, if client wishes to use new building before completion certified

尊敬的先生：

我方得知，你方希望在我方证明实际竣工［使用 GC/Works/1（1998）合同时，为“工程按合同竣工”］之前使用该建筑物/建筑物的一部分［视情况取舍］。

我方强烈建议你方不要这样做，因为这样可能会引起某些复杂情况。如果你方一定要使用该建筑物，承包商可能会声称因你方提前使用而造成了损坏，并应对此负责。这种声称很难完全被驳倒。

此事最终由你方来决定，请给予指示。这样做必须获得承包商的许可，且涉及到的保险事项必须与你方的保险经纪人进行讨论。

你忠诚的

Dear Sir

I understand that you wish to use the building/part of the building [*delete as appropriate*] before I certify practical completion [*substitute* ‘*that the Works are completed in accordance with the contract*’ *when using GC/Works/1* (*1998*)].

I strongly advise you not to follow this course of action, because it may give rise to complications. If it is absolutely essential that you use the building, the contractor may well assert that such use has caused damage for which he will hold you liable. Such assertions are always difficult to refute completely.

It is a matter for you to decide and let me have your instructions. The contractor's permission must be obtained and there are insurance implications to discuss with your insurance broker.

Yours faithfully

函件 227

竣工之前，致委托人

Letter 227

To client, prior to completion

尊敬的先生：

预计该建筑将于［填入日期］竣工并交付你方使用。我方打算安排你方与承包商会面，以便当天交接钥匙。如果你方能抽出当天下午的时间，一旦我方确定了不会再有临时变故，我方将确认此安排。

工程交接后，承包商不再承担保险义务，你方应该安排全额保险，从［填入日期］起生效。为此，我方帮你方估算了一下，建筑的造价为［填入金额］，但你方应该向你方的保险经纪人咨询必要的事项，以保证你方的保险包括了重建成本和专业费用。

你忠诚的

抄送：工程监督员

Dear Sir

I anticipate that the building will be ready for you to take possession on [*insert date*] and I intend to arrange a meeting with the contractor so that you can accept the keys on that date. If you will provisionally reserve the afternoon of that day, I will confirm arrangements as soon as I am satisfied that there are unlikely to be any last minute hitches.

The contractor's insurance obligations will cease on handover and you should make arrangements for full insurance cover effective from [*insert date*]. In order to assist you, I estimate the building value at [*insert amount*], but you should consult your insurance broker regarding the need to ensure that your insurance includes all rebuilding costs and professional fees.

Yours faithfully

Copy: Clerk of works

函件 228
致委托人，确认工程交接会议

Letter 228
To client，confirming handover meeting

尊敬的先生：

我方已于［填入日期］去函，告知你方希望能于［填入日期］下午举行工程交接会议。

［如果日期已经确认，可以加上：］

现欣然确认此安排，我方提议于［填入时间和其他必须的相关事宜，如午餐安排］去接你方。

［如果日期变更，改为：］

非常遗憾，工程的某些重要部分不能按期完工，因此我方建议于［填入日期］举行工程交接会议。如果方便的话，我方将于［填入时间和其他必需的相关事宜，如午餐安排］去接你方。

［然后：］

请告知此日期和时间是否方便。

你忠诚的

Dear Sir

I refer to my letter of the [*insert date*] in which I said that I expected to be able to hold a handover meeting in the afternoon of [*insert date*].

[*Add, if date confirmed:*]

I am pleased to be able to confirm this arrangement and I propose collecting you at [*insert time and any further matters, such as lunch, which you have to settle*].

[*Add, if date changed:*]

Unfortunately certain important parts of the work will be unfinished on that day and I, therefore, suggest [*insert date*] for the handover meeting. If this is convenient, I propose collecting you at [*insert time and any further matters, such as lunch, which you have to settle*].

[*Then:*]

Please let me know that the date and time are convenient.

Yours faithfully

函件 229

关于竣工前的检查，致承包商

Letter 229

To contractor regarding inspection before completion

尊敬的先生：

工程交接会议将于［填入日期］举行，委托人到时会到场。我方建议在此之前，于［填入日期］对工程进行全面检查。

所有顾问都将到场。欣然希望你方能做好一切准备工作，准备好所有钥匙，并安排［填入承包商代表姓名］参加。

你忠诚的

抄送：工料测量师
顾问
工程监督员

Dear Sir

I propose to visit site on [*insert date*] to carry out a full inspection of the Works before the handover meeting on the [*insert date*] at which the client will be present.

All consultants will be in attendance and I should be pleased if you would arrange to have everything ready, all keys available and [*insert name of contractor's representative*] on hand.

Yours faithfully

Copies: Quantity surveyor
Consultants
Clerk of works

函件 230

关于竣工前的检查，致顾问

Letter 230

To consultants, regarding inspection before completion

尊敬的先生：

随函附上我方致承包商的信函副本，此信函内容不释自明。

请确认你方将出席会议。

你忠诚的

Dear Sir

I enclose a copy of my letter to the contractor which is self-explanatory.

Please confirm that you will be present at the meeting.

Yours faithfully

函件 231

工程交接后，致委托人

Letter 231

To client, after handover

尊敬的先生：

谨提及在工地举行的由［填入姓名和公司名称］参加的以上项目的工程交接会议。现确认你方在检查建筑物之后表示总体满意，并接受了［填入接受的钥匙、文件等的细节表］。你方自行投保的保险现在应该已经生效。如果你方的保险经纪人没有对此确认，请立即要求其确认。

缺陷责任期［使用 GC/Works/1（1998）合同时，为“维修期”］从实际完工［使用 GC/Works/1（1998）合同时，为“按合同规定完工”］持续到［填入日期］。尽管我方将在此期间进行检查，如果你方能记录下你方注意到的缺陷，以便我方将其列入缺陷清单中，将会非常有帮助。如果有任何明显的工程缺陷引起你方的不便，请立即告知我方，以便我方指示承包商立即修缮。

你忠诚的

Dear Sir

I refer to the handover meeting held on site at the above project at which [*list names and firms*] were present. I confirm that after an inspection of the building you expressed yourself generally satisfied and accepted [*insert detailed list of all keys, documents, etc. accepted*]. Your own insurance should now be effective. If your broker has not confirmed it, press him to confirm without delay.

The defects liability [*substitute 'maintenance' when using GC/Works/1 (1998)*] period extends from the date of practical completion [*substitute 'completion in accordance with the contract' when using GC/Works/1 (1998)*] to [*insert date*]. Although I will make my own inspection during this period, it would be helpful if you would make a note of any defects which you notice so that I can include them on my list. If any defects become apparent and cause you any inconvenience, please let me know so that I can instruct the contractor to attend to them immediately.

Yours faithfully

函件 232

如果寄发缺陷清单，致承包商

Letter 232

To contractor, if sending schedule of defects

尊敬的先生：

缺陷责任期［使用 GC/Works/1（1998）合同时，为“维修期”］截止于［填入日期］。按合同第 17.2 条［使用 WCD98 合同时，替换为“第 16.2 条”；使用 IFC98 合同时，替换为“第 2.10 条”；使用 MW98 合同时，替换为“第 2.5 条”；使用 GC/Works/1（1998）合同时，替换为“第 21 条”］，现随函附上我方于［填入日期］进行工地检查时发现的工程缺陷的清单。欣然希望你方能立即关注这些缺陷。

你忠诚的

抄送：工程监督员

Dear Sir

The defects liability [*substitute* '*maintenance*' *when using GC/Works/1* (*1998*)] period ended on the [*insert date*]. In accordance with clause 17.2 [*substitute* '*16.2*' *when using WCD 98*, '*2.10*' *when using IFC 98*, '*2.5*' *when using MW 98 or* '*21*' *when using GC/Works/1* (*1998*)] of the contract I enclose a schedule of the defects I found during my inspection carried out on the [*insert date*]. I should be pleased if you would give your immediate attention to these defects.

Yours faithfully

Copy: Clerk of works

函件 233

如果在缺陷责任期内需要立即注意，致承包商

此信函仅适用于 JCT 98 合同或 WCD 98 合同

Letter 233

To contractor if immediate attention required during the defects liability period

This letter is only suitable for use with JCT 98 or WCD 98

尊敬的先生：

尽管有合同第 17.2 条［使用 WCD98 合同时，替换为“第 16.2 条”］各项的说明，依据第 17.3 条［使用 WCD98 合同时，替换为“第 16.3 条”］，我方有权在我方认为有必要时，发出工程缺陷整改指示。

随函附上的整改指示即属此类，十分希望你方能执行此指示。

你忠诚的

抄送：工程监督员

Dear Sir

Notwithstanding the provisions of clause 17.2 [*substitute '16.2' when using WCD 98*] of the contract, clause 17.3 [*substitute '16.3' when using WCD 98*] empowers me to issue instructions regarding the making good of defects whenever I consider it necessary to do so.

The enclosed instruction refers to making good which falls into this category and I should be pleased if you would carry out the instruction forthwith.

Yours faithfully

Copy: Clerk of works

函件 234

致承包商，在缺陷责任期/维修期内要求整改

Letter 234

To contractor, requiring making good during the defects liability/maintenance period

尊敬的先生：

请紧急进行附件指示中的整改工作。

此指示是按第 17.3 条［使用 WCD 98 合同时，替换为“第 16.3 条”；使用 IFC98 合同时，替换为“第 2.10 条”；使用 MW98 合同时，替换为“第 2.5 条”；使用 GC/Works/1（1998）合同时，替换为“第 40(2)(j)条”］发出的。

你忠诚的

抄送：工程监督员

Dear Sir

As a matter of urgency, please carry out making good as indicated on the enclosed instruction.

The instruction is issued in accordance with clause 17.3 [*substitute '16.3' when using WCD 98, '2.10' when using IFC 98, '2.5' when using MW98 or '40(2)(j)' when using GC/Works/1 (1998)*].

Yours faithfully

Copy: Clerk of works

函件 235

如果某些缺陷不进行整改，致委托人

此信函不适用于 GC/Works/1（1998）合同

Letter 235

To client, if some defects are not to be made good

This letter is not suitable for use with GC/Works/1（1998）

尊敬的先生：

我方已明确你方不要求承包商对以下缺陷进行整改：[列出]

这此缺陷已经包括在缺陷责任期末我方向承包商发出的缺陷列表中。欣然希望你方确认以下事项，以便我方发出合适的指示。

1. 你方不要求承包商对此信函中列出的缺陷进行整改。

2. 你方授权我方从合同总额中进行合理的扣除。

3. 你方放弃就上述缺陷清单中列出的且未进行整改的缺陷追究任何人责任的权利。

4. 如果第三方关于这些缺陷提出索赔，你方同意免除我方的责任。

你忠诚的

Dear Sir

I understand that you do not require the contractor to make good the following defects：[*list*].

These defects are included in my schedule of defects issued to the contractor at the end of the defects liability period. In order that I may issue the appropriate instructions I should be pleased if you would confirm the following：

1. You do not require the contractor to carry out making good to the defects listed in this letter.

2. You authorise me to make an appropriate deduction from the contract sum.

3. You waive any rights you may have against any persons in regard to the items listed as defects in the above-mentioned schedule of defects and not made good.

4. You agree to indemnify me against any claims made by third parties in respect of such defects.

Yours faithfully

函件 236

指示某些缺陷不需要进行整改，致承包商

此信函不适用于 GC/Works/1（1998）合同

Letter 236

To contractor, instructing that some defects are not to be made good

This letter is not suitable for use with GC/Works/1（1998）

尊敬的先生：

缺陷责任期将于［填入日期］结束。我方于［填入日期］对工程进行了检查，并附上了所发现的缺陷的清单。

依据合同第 17.2 条［使用 WCD98 合同时，替换为“第 16.2 条”；使用 IFC98 合同时，替换为“第 2.10 条”；使用 MW98 合同时，替换为“第 2.5 条”］，我方特此发出指示，你方不需要对清单中列出的/标有“E”的［视情况取舍］缺陷进行整改。

关于不要求你方进行整改的缺陷，我方将从合同总额中扣除合理的金额。

你忠诚的

抄送：雇主
　　　工料测量师
　　　工程监督员

Dear Sir

The defects liability period ended on the [*insert date*]. I inspected the Works on the [*insert date*] and enclosed is a schedule of the defects found.

I hereby instruct that, in accordance with clause 17.2 [*substitute '16.2' when using WCD 98, '2.10' when using IFC 98 or '2.5' when using MW98*], you are not required to make good any of the defects shown on the schedule/those defects marked 'E' [*delete as appropriate*].

An appropriate deduction will be made from the contract sum in respect of the defects which you are not required to make good.

Yours faithfully

Copies: Employer
Quantity surveyor
Clerk of works

函件 237

致承包商，索要竣工记录

Letter 237

To contractor, requiring ‘as-built’ records

尊敬的先生：

一份建筑物的完整、准确的竣工记录对雇主将来的维修程序是十分必要的。合同文件［若在缺陷责任期开始之前仍然没有收到记录，则标明工程量清单或工程详细说明中的相应页码和参考条款，当使用 WCD98 合同时，参照“第 5.5 条”］中也需要这些记录。

考虑到你方记录涉及的范围，你方可能需要联系相应的分包商并采取额外措施。我方必须强调，你方对此记录的完整性和准确性负有责任。

请于下周内通知我方何时可以收到一套完整的竣工记录。

你忠诚的

Dear Sir

A complete and accurate record of the building as built is essential for the employer's future maintenance procedures. Such records are required in the contract documents [*indicate position in bills of quantities or specification by stating the appropriate page numbers and references, refer to ‘clause 5.5’ when using WCD 98 if the records were not received before the commencement of the defects liability period*].

Depending on the extent of your own records, you may be involved in contacting the appropriate sub-contractors and taking additional measurements. I must stress that completeness and accuracy of such records is entirely your responsibility.

Please inform me, during the next week, when I can expect to receive a full set of as-built records.

Yours faithfully

函件 238

致承包商，要求归还所有的工程图纸和文件材料

此信函仅适用于 JCT98 合同

Letter 238

To contractor, requiring return of all drawings and documents

This letter is only suitable for use with JCT 98

尊敬的先生：

依据合同第 5.6 条，我方正式要求你方归还我方所有的工程图纸、详细资料、说明表、工程量清单或工程详细说明，以及其他标有我方名字的同类文件资料。

你忠诚的

Dear Sir

In accordance with clause 5.6 of the contract I formally request you to return to me all drawings, details, descriptive schedules, bills of quantities or specification or other like documents which bear my name.

Yours faithfully

函件 239

致承包商，要求归还所有的工程图纸和文件材料

Letter 239

To contractor, requiring return of all drawings and documents

尊敬的先生：

请你方归还我方所有的工程图纸、详细资料、说明表、工程量清单或工程详细说明，以及其他标有我方名字的同类文件资料。

这些文件资料版权属我方所有。未经我方明确的书面许可，这些文件及其内容不可用于任何目的。

你忠诚的

抄送：雇主
工料测量师
顾问
工程监督员

Dear Sir

Please return to me all drawings, details, schedules and other like documents which bear my name.

The documents are my copyright and neither they nor the information they contain may be used for any purpose whatsoever without my express written permission.

Yours faithfully

Copies: Employer
Quantity surveyor
Consultants
Clerk of works

第十二章　反馈

建筑师们对于工程结束时征询反馈信息的意义持不同意见。大多数建筑师同意，从理论上来说调查反馈信息是好事，但实际上极少有人全心全意地进行征询。多数人不情愿这样做的原因主要是因为害怕委托人可能会借建筑师征询反馈意见的机会提出索赔要求。由于永远存在着诉讼的威胁，这并不奇怪。函件244和245就是用来应对隐蔽缺陷暴露出来时的情况。

函件 240

致委托人，征询反馈信息

Letter 240

To client, requesting feedback information

尊敬的先生：

我方总是尽力在建筑物投入使用几个月后进行一次反馈信息调查活动。我方发现这样做既有助于解决任何自然暴露出来的问题，也有助于我方在以后的工作中不断地优化服务程序。

我方正在安排与设计团队成员进行一次会议。如果你方能告知有空出席会议的日期，我方将不胜感激。

会上要讨论的议题包括：工程全过程中沟通程序的有效性以及竣工后建筑运作的满意情况。

你忠诚的

Dear Sir

I always endeavour to carry out a feedback exercise after the building has been in use for a few months. I have found it useful for resolving any problems which may present themselves and to assist in refining my procedures for future work.

I am arranging a meeting with members of the design team and I should be grateful if you could let me have some dates on which vou would be free to attend.

Among the subjects for discussion will be the effectiveness of communication procedures throughout all stages of the work and the satisfactory operation of the finished building.

Yours faithfully

函件 241

致顾问，征询反馈信息

Letter 241

To consultants, requesting feedback information

尊敬的先生：

我方总是尽力在建筑物投入使用几个月后进行一次反馈信息调查活动。我方发现这样做既有助于解决任何自然暴露出来的问题，也有助于改进服务程序使双方受益。如果将来能再次合作，这将尤其有用。

我方正在安排与设计团队成员进行一次会议，并邀请了委托人参会。

会上要讨论的议题包括：工程全过程中沟通程序的有效性以及竣工后建筑运作的满意情况。

如果你方能告知几个你方有空出席会议的日期，我方将不胜感激。

你忠诚的

Dear Sir

I always endeavour to carry out a feedback exercise after the building has been in use for a few months. I have found it useful for resolving any problems which may present themselves and to assist in improving procedures to our mutual benefit. This could be especially useful in the event of future work together.

I am arranging a meeting with all the members of the design team and I have invited the client to attend and to contribute to the discussion.

Among subjects for discussion will be the effectiveness of communication procedures throughout all stages of the work and the satisfactory operation of the finished building.

I should be grateful if you would let me have some dates on which you are free to attend.

Yours faithfully

函件 242

致承包商，征询反馈信息

Letter 242

To contractor, requesting feedback information

尊敬的先生：

我方总是尽力在建筑物投入使用几个月后进行一次反馈信息调查活动。我方发现这样做既有助于解决任何自然暴露出来的问题，也有助于改进服务程序使双方受益。如果将来能再次合作，这将尤其有用。

我方正在安排与设计团队成员进行一次会议。如果你方能告知几个你方有空出席会议的日期，我方将不胜感激。

会上要讨论的议题包括：工程全过程中沟通程序的有效性以及竣工后建筑运作的满意情况。

你忠诚的

Dear Sir

I always endeavour to carry out a feedback exercise after the building has been in use for a few months. I have found it useful for resolving any problems which may present themselves and to assist in improving procedures to our mutual benefit. This could be especially useful in the event of future work together.

I am arranging a meeting with all the members of the design team and I should be grateful if you would let me have some dates on which you are free to attend.

Among subjects for discussion will be the effectiveness of communication procedures throughout all stages of the work and the satisfactory operation of the finished building.

Yours faithfully

函件 243

如果在缺陷责任期过后，委托人要求建筑师检查疑似的缺陷，致委托人

Letter 243

To client, if architect asked to inspect suspected defects after the end of the defects liability period

尊敬的先生：

谨提及你方于［填入日期］的电话/来函［视情况取舍］要求我方对工程的一个疑似缺陷进行检查。

正如你方所知缺陷责任期已过，我方已经向承包商发出了缺陷清单。你方所指的疑似缺陷在我方检查时并未发现。除了核查承包商已完成的所有指出的缺陷整改，并对此签发证明以及最终证明之外，我方的服务已经完成。当然，缺陷责任期结束并不意味着承包商潜在责任的完结。承包商仍然要对不符合合同要求的工程或材料负责。因为，这些情况属于承包商违约。

如果你方希望的话，我方很乐意对疑似的缺陷进行检查。我方对此项工作将按目前仍有效的雇佣条款的规定收费，费用标准为每小时［填入费率］，再加上其他支出费用［如果需要，加上“再加增值税”］。请确认是否同意我方的收费。

你忠诚的

Dear Sir

I refer to your telephone call/letter [*delete as appropriate*] of the [*insert date*] when you asked me to investigate a suspected defect in the Works.

As you know, the defects liability period has ended and I have issued a schedule of defects to the contractor. The suspected defect to which you refer was not apparent during my inspection. Other than checking to see that the contractor has completed the making good of all defects notified and issuing a certificate to that effect and, eventually, the final certificate, my services are complete. Of course, the end of the defects liability period does not signal the end of the contractor's potential liability. The contractor is still liable for work or materials which are not in accordance with the contract. They are breaches of contract on his part.

I would be happy to visit site and to investigate the suspected defect if you so wish. My fee for carrying out this work will be charged at the rate of £ [*insert rate*] per hour plus expenses [*add*: '*plus VAT*' *ifapplicable*] under the same terms of engagement currently in force. Please confirm your agreement to my charge.

Yours faithfully

函件 244

关于发出最终证明后的隐蔽缺陷，致委托人

Letter 244

To client regarding latent defects after the final certificate

尊敬的先生：

从你方［填入日期］的来函/电话［视情况取舍］中，我方得知，你方认为出现了一个隐蔽缺陷并为此担忧。

［如果缺陷责任诉讼时效可能将到期，可以加上：］

由于此问题的缺陷责任诉讼时效将很快到期，为保障你方的权利，我方强烈希望你方立即寻求法律建议。目前我方不可能对责任进行确定，但完全有理由考虑前承包商与此的关系。如果你方希望我方处理此事，我方乐意帮忙。现随函附上 RIBA 雇佣条款 SFA/99［视情况替换为“CE/99”或“SW/99”］以及我方的收费和开支细节，请你方同意。

［否则，如果缺陷责任诉讼时效尚未到期，可以加上：］

你方完全有理由考虑前承包商与此缺陷的关系。如果你方希望我方处理此事，我方乐意帮忙。现附上 RIBA 雇佣条款 SFA/99［视情况替换为“CE/99”或“SW/99”］以及我方的收费和开支细节，请你方同意。

你忠诚的

Dear Sir

Following your letter/telephone call [*delete as appropriate*] of the [*insert date*], I understand that you are concerned about what you believe to be a latent defect.

[*If there is any possibility that the limitation period for the defect may expire add*:]

The limitation period in respect of this problem may soon expire. In order to safeguard your rights, I strongly urge you to seek legal advice on the point immediately. It is not possible for me to take a view on liability at this stage, but it makes sense to involve the original contractor. If you wish me to deal with this matter, I would be happy to assist and a copy of the RIBA terms of engagement SFA/99 [*substitute 'CE/99' or 'SW/99' as appropriate*], together with details of my fees and expenses are enclosed for your agreement.

[*Otherwise, add*:]

It makes sense to involve the original contractor. I shall be happy to deal with this matter if you wish and a copy of the RIBA terms of engagement SFA/99 [*substitute 'CE/99' or 'SW/99' as appropriate*], together with details of my fees and expenses are enclosed for your agreement.

Yours faithfully

函件 245

关于隐蔽缺陷，致承包商

Letter 245

To contractor, regarding latent defects

尊敬的先生：

我方的委托人要求我方处理以上工程于［填入日期］出现的一个缺陷。该问题在于［简要描述缺陷］。

我方进行了初步检查，我方认为此缺陷是你方的责任。请于本周内电话联系我方，以便安排共同检查。

你忠诚的

Dear Sir

My client asked me to deal with a defect in the above Works which became apparent on or about [*insert date*]. The problem appears to be [*briefly describe the defect*].

I have carried out a preliminary inspection and my opinion is that the defect is your responsibility. Please telephone me during the week in order to arrange a joint inspection.

Yours faithfully

词汇对照

Abortive work	窝工
Adjudication	裁决
Agent	代理人
Amenity societies	市容协会
Analysis	分析
Appraisal	评估
Arbitration	仲裁
Architect's instructions	建筑师指示
Architect, other	其他建筑师
As-built records	竣工记录
Authorized representatives	授权代表
Bills of quantities	工程量清单
Boundaries	界限
Brochure	作品宣传册
Building Regulations	建筑规范
Certificate	证明书
Financial certificate	财务证明
General certificate	一般证明
Non-completion certificate	未竣工证明
Clause	条款
Clean Air Act	空气清洁法案
Clerk of works	工程监督员
Client	委托人
Coal Authority	煤炭管理部门
Coal Mining Subsidence Act 1991	1991 年煤矿津贴法案
Code of Procedure for Selective Tendering for Design and Build 1996	1996 年设计施工招投标程序规范
Code of Procedure for Single Stage Selective Tendering 1996	1996 年单一阶段招投标程序规范

Code of Procedure for Two Stage Selective Tendering 1996
1996 年两段招投标程序规范
Code of practice 惯例
Code of professional conduct 职业操守规范
Common law claims 依据普通法的索赔
Completion 竣工
Conditions of Appointment for Small Works 1999
1999 年小型建筑建筑师委任协议
Conditions of Engagement for the Appointment of an Architect 1999
1999 年中型工程建筑师委任协议
Construction Contracts（Northern Ireland）Order 1997
1997 年（北爱尔兰）建筑合同法令
Consultants 顾问
Consumer 客户
Contract 合同
Contract documents 合同文件
Contract Sum Analysis 合同总额分析
Contractor client 承包商委托人
Contractor's proposals 承包商建议书
Contractor's Statement 承包商报表
Copyright 著作权
Correspondence 信函
Costs 造价

Decisions 决定
Deed 契约
Default notice 违约通知
Defects 缺陷
Defects liability period 缺陷责任期限
Deferment of possession 工地进驻延迟
Design 设计
Design team 设计团队
Detailed proposals 详细建筑方案
Determination 终止
Direction 指示
Discrepancies 矛盾
Divergence 分歧
Drainage authority 排污管理部门
Drawings 工程图纸

Easements	地役权
Electricity supplier	电力供应商
Employer's agent	雇主代理人
Employer's Requirements	雇主要求
Environmental health	环境健康
Environmental service	环境服务
Ex gratia claims	通融索赔
Expense	费用开支
Extensions of time	工程延期
Further	再次延期
General	一般延期
Given	特定延期
In parts	部分延期
Nominated sub-contract	指定分包合同延期
No grounds	无依据延期
Not due	不应当延期
Report	延期报告
Review	延期审查
Failure of work	工程不合格
Feasibility	可行性
Fees	费用
Accounts	费用账单
Additional	额外费用
Agreement	费用协议
General	一般费用
Payment in advance	预付费用
Tender	投标费用
Final proposal	最终建议书
Final statement	最终报表
Fire authority	消防管理部门
Fire certificate	消防证书
Fire Precautions Act 1971	1971 年消防法案
Fire prevention officer	消防官员
Gas supplier	燃气供应商
Geotechnical specialist	地质专家
Gifs	礼物
Ground investigation	地质勘测

Handover meeting 交接会议
Health and Safety Executive 健康与安全管理人员
Highway Authority 公路管理部门
Housing Grants, Construction and Regeneration 1996
1996 年住房补贴、建造和改造法案

Information 信息
Inspection 检查
Insurance 保险
Contractor 承包商保险
Employer 雇主保险
Professional indemnity 专业责任保险
Interest 利息

Latent defects 隐蔽缺陷
Legal proceedings 法律程序
Letter of intent 合同意向书
Liability 责任
Liquidated damages 预定违约金
Local authority 当地政府部门
Loss and/or expense 损失和（或）费用
Acceptance 损失和（或）费用接受
Ascertainment 损失和（或）费用确定
Badly presented 损失和（或）费用表述不清
Further information 损失和（或）费用补充信息
General 一般损失和（或）费用
Rejection 损失和（或）费用拒付
Report 损失和（或）费用报告
Supplementary provisions 损失和（或）费用补充条款

Maintenance period 维修期
Manufacturer 制造商
Master programme 总进度计划
Materials 材料
Minutes 会议记录

Named persons 指定人员
Contractors 指定承包商
New materials and processes 新材料和工艺

Nominated sub-contractors	指定分包商
Extension of time	分包商工程延期
Failure to complete	分包商工程不合格
General	一般分包商
Payment	分包商付款
Notice	通知
Objections	反对
Off-site materials bond	场外材料保证书
Opening up	打开
Operations on site	施工现场活动
Outline proposals	初步建筑方案
Ownership	所有权
Payment	付款
Performance bond	履约保证书
Performance specified work	规定实施工程
Planning	规划
Advertisement	规划公告
Applications	规划申请
Approvals	规划许可
Authority	规划管理部门
Fees	规划费用
Full approval	完全规划许可
Outline approval	初步规划许可
Proceeding before approval	规划许可之前进行施工
Reserved matters	规划保留事项
Temporary permission	临时规划许可
Plant testing and commissioning	设施测试和调试
Possession of site	工地进驻
Power	电力
Practical completion	实际竣工
Priced activity schedule	标价分项工程表
Priced statement	标价报表
Production information	生产信息
Project planning	项目规划
Progress	进度
Provisional sums	预算金额

Quantity surveyor 工料测量师

Referee 公断人
Referral 相关事项
Refuse collection 废物收集
Relevant events 相关事项
Restrictive covenant 限制契约

Samples 样品
Schedule of work 工程明细表
Sectional completion 分段竣工
Services 服务
Setting out 放线
Site 工地
Specification 工程详细说明
Standard Form of Agreement for the Appointment of an Architect 1999
1999 年建筑师委任协议标准格式
Statutory requirements 法定要求
Strategic brief 基本设计要点
Sub-contractors 分包商
Sub-letting 分包
Supplier 供应商
Survey 调查
Suspension of architect's work 建筑师工作暂停
Suspension of contractor's work 承包商工作暂停

Technical literature 技术说明文字材料
Technical representative 技术代表
Telephone service supplier 电话服务供应商
Tender 投标
 Contractor 承包商投标
 General 总投标
 Sub-contractor 分包商投标
Termination of appointment 委任关系终止
Terms of engagement 雇佣条款
Title deeds 地契

Unfair Contract Terms in Consumer Contracts Regulations 1999
1999 年客户合同中的不公平条款规定

Valuations	估价
Variations	变更
Water supplier	自来水供应商
Warranty	担保书
Withholding notice	扣除通知
Workmanship	工艺

作者简介

大卫·查贝尔（拥有建筑学荣誉学士、建筑学硕士、法学硕士和博士学位，是RIBA会员）。曾任多家公共及私人机构的建筑师以及建筑承包商的合同管理者。现任一家建筑合同咨询公司——查贝尔－马修有限公司的董事长，并经常担任裁决人和仲裁人。1996年1月起，他被任命为贝尔法斯特女王大学的访问教授，并于2000年1月被授予建筑执业与管理研究学科的资深研究员资格且获得教授头衔。2003年1月起，他担任伯明翰中英格兰大学的建筑执业管理和法律学科的访问教授。查贝尔先生在建筑行业界的专业刊物上发表了很多文章，并为业界编著了20余本书籍。同时，他还作为RIBA的专家顾问之一经常在各地讲学。